A SHEARWATER BOOK

Making *Sense* of Sex

Making *Sense* of Sex

HOW GENES *and* GENDER
INFLUENCE *our* RELATIONSHIPS

DAVID P. BARASH, PH.D. AND
JUDITH EVE LIPTON, M.D.

ISLAND PRESS / Shearwater Books
Washington, D.C. • Covelo, California

A Shearwater Book
published by Island Press

Copyright © 1997 David P. Barash and Judith Eve Lipton

All rights reserved under International and Pan-American Copyright Conventions. No part of this book may be reproduced in any form or by any means without permission in writing from the publisher: Island Press, 1718 Connecticut Avenue, N.W., Suite 300, Washington, DC 20009.

Shearwater Books is a trademark of The Center for Resource Economics.

Library of Congress Cataloging-in-Publication Data

Barash, David P.
    Making sense of sex: how genes and gender influence our
    relationships / David P. Barash, Judith Eve Lipton.
        p.   cm.
    Includes bibliographical references and index.
    ISBN 1-55963-452-9 (cloth)
    1. Sex. 2. Sex (Biology)—Evolution. 3. Sex differences.
    4. Human evolution. 5. Social evolution.   I. Lipton, Judith, Eve.
    II. Title.
    GN235.B37   1997
    306.7—dc21   97-25746
                                                            CIP

Printed on recycled, acid-free paper ✪

Manufactured in the United States of America

10 9 8 7 6 5 4 3 2 1

Contents

Preface

THIS BOOK IS ABOUT MALES AND FEMALES, men and women, girls and boys, and why the two sexes are different. Most people understand the basic physical differences, and several popular books have described differences in how males and females communicate, but here we explore the origins of sex itself. We then look at its consequences, including lust, parenting, and childhood—in other words, the fundamental aspects of life for most people. In making sense of sex and all its trappings, we rely on the work of many biologists who, over the years, have come up with a simple, wonderful, and even beautiful explanation of why men and women are different, from genes to bodies to behavior. Our aim is to describe and explain this general theory and then discuss its implications. In the end, we hope our readers will come away from this book understanding themselves and their relationships in a new and more meaningful way.

In encountering the ideas in this book, it may help the reader to understand who we, the authors, are and what drives us professionally and personally. David Barash is a middle-aged evolutionary biologist and professor of psychology at the University of Washington who specializes in animal behavior, reveling in the study of obscure species for what they reveal about larger evolutionary issues. Judith Lipton is a psychiatrist in private practice with an abiding fondness for music and horses. She's someone who understands how the brain works, what to do about one's mother, and how to make peace with a former spouse. Each of us was previously married, and each has children from the previous marriage. To our blended family, Judith brought a son and David brought a daughter; together, we had two daughters.

We consider ourselves equals: intellectually, educationally, and financially. Each of us has a doctoral degree (David's in biology, Judith's in medicine), each of us has a successful career, and each spends considerable time with our children. Still, we find it impossible to detach ourselves from gender issues in our own lives, from courtship, marriage, and divorce to plain old sexual relations and, then, parenting and step-parenting. We consider ourselves to be as nonsexist a family as one could comfortably imagine, and as a couple we are deeply in love as well as avowedly monogamous. And yet, even after twenty years of marriage and years of professional training and practice, we find ourselves struggling to make sense of our own gender gap, ranging from modes of verbal expression to sexual proclivities. We say this to emphasize the broad nature of what we mean by the terms *sex* and *sex differences*, referring not just to genitals but to hormonal, physical, and behavioral differences throughout the life cycle, differences that deeply affect each of us as well as every other human being.

Even the process of collaboration has made us aware (sometimes painfully so) of our own differences. Left to his own, David would present the material in this book as if it were a lecture, intellectually interesting and fun but somewhat detached. Judith wants to relate to her readers and so tends to be more empathic and emotional, sympathetic and resonant; she wrote everything that is self-revealing in this book.. She also prefers to present case histories and offer therapeutic suggestions and interpretations, whereas David prefers to discuss theory or animals. Overall, David likes to discourse on general principles. The patterns of adultery and violence he describes on these pages, for example, are the practices and patterns of animal and human societies all over the world. It is not that David can't be empathic or intuitive or self-disclosing or that Judith can't quote data and abstract theory or maintain interpersonal boundaries; rather, we tend to look at life differently, and some of that difference exists because we are male and female.

We also experience work differently. In his studies of animal behavior, David has spent thousands of hours with binoculars watching species as varied as bluebirds and bears, trying to be as unobtrusive as possible. His working concept of an experiment rarely interferes with the lives of his subjects, and he seeks to build general principles out of specific cases. In contrast, Judith as a clinical psychiatrist, works with individual patients, confronting such problems as adultery and domestic violence in the context of individual lives and families. She views each new patient as an experiment about to unfold, a relationship in the making. Her goal as a clinician *is* to interfere, to have a positive impact on someone else's life. Her thinking tends to be the inverse of David's: she begins with general principles of psychiatry and medicine and applies them to the individual circumstances of each patient.

When, on the lecture circuit, a questioner disagrees with us, Judith tends to think, "We must have been unclear in our presentation," whereas David is likely to wonder why the questioner is so obtuse or, if persistent, such a jerk. Judith tends to question, David to argue. On his own, David would make this a popular science book; on her own, Judith would write a self-help book. We find it a rather humorous irony that despite our scholarly degrees and feminist politics, our actual behaviors fall into rather predictable stereotypes. Maybe our joint effort will both amuse and console.

As a husband–wife team, we have brought our own perspectives to the subject at hand and are convinced that by collaborating in this way, we can offer our readers a more creative synthesis and a more enlightening view of gender differences than would be possible had either one of us authored this volume alone. The upshot is a controversial view of human beings and their sexuality. We gratefully acknowledge our reliance on the pioneering work of many biologists, some of whom worked mainly in the realm of theory and others who spent many hundreds of hours observing the behavior of animals as well as human beings. Sometimes we cite them by name in the text; at other times, in the endnotes. In nearly all cases, they did the scientific heavy lifting. We merely point out connections and—we hope—convey the results in an accessible way to our readers, nearly all of whom we assume are non-scientists.

Some readers will dismiss this approach outright because it gives a scientific—moreover, a Darwinian—rather than a theological explanation for human nature. Anyone who believes, perhaps for religious reasons, that biology and animal behavior have nothing to do with our own species might as well stop here. We are no more inclined to dispute evolution versus creationism than we are inclined to argue with members of the Flat Earth Society. Instead, we present a fresh interpretation of sex and sex differences based on evolutionary biology, and we assume our readers are open-minded and curious enough to give it a try.

We have included some case histories to emphasize that the issues we discuss are not purely theoretical and to demonstrate how biology can help make sense of life's day-to-day experiences. These cases are true in that each one happened to a real person, but names have, of course, been changed, along with minor details, and often several stories have been amalgamated into one. Although we make some value judgments—about violence, for example—they are clearly our own, and we do not presume to tell anyone what to do or how to be. Indeed, we expect that many of our readers will disagree from time to time with what we say. Our hope is that the facts and theories we present will help each reader of this book to choose his or her course in life, armed with insights and information and prepared to live and let live with all the freedom human beings can muster.

DAVID P. BARASH, PH.D.
JUDITH EVE LIPTON, M.D.

Acknowledgments

No book, it is said, is written alone, and *Making Sense of Sex* is no exception. Most of all, we would like to thank . . . each other! This book would not have happened without the two us. It was written as improvisational jazz is performed: each of us contributed themes, variations, assorted harmonies, riffs, and rhythms.

We owe numerous debts outside our private *folie à deux*. We especially thank the many biologists—too many to list individually here—whose theoretical insights and old-fashioned hard work helped reveal the underlying principles as well as the basic "facts of life" on which our presentation is based. Similarly, we thank the anthropologists, psychologists, and other social scientists whose evolutionary orientation has propelled them against the grain of traditional wisdom in their fields. We also thank Judith's patients and David's students for providing source material and valuable responses.

A number of people read, commented on, and in various ways improved this book. These include Carla Bradshaw, Leif Carlson, Daniel M. T. Fessler, Sarah Kinney, Kate Noble, Adam Sorscher, David Stutz, and Del Thiessen, as well as two anonymous reviewers recruited by Island Press. We are very grateful to them all.

Our editor, Laurie Burnham, believed in this project, stuck with it through good times and bad, and was extraordinarily attentive to all aspects of its development, in the process challenging, delighting, enlightening, and infuriating us—sometimes all at once! Thank you, Laurie, for hanging in there, working so hard, and contributing so much.

We gratefully acknowledge the many animals of Meadowland Farm for teaching us about practical aspects of sex differences.

Our daughters, Nellie, Ilona, and Eva Barash were wonderfully tolerant—even helpful—throughout this process. Ilona and Eva were founts of popular culture who kept us suitably humble. Nellie provided distraction and comic relief. We thank them for the inspiration they provided and for projecting our genes into the future.

Many thanks are due our readers, as well, for trying on this new perspective on sexuality, evolution, and relationships. We hope they will tolerate our sassiness and enjoy themselves.

CHAPTER 1

Differences

THERE IS GRANDEUR in this view of life. . . . Whilst this planet has gone cycling on according to the fixed law of gravity, from so simple a beginning endless forms most beautiful and most wonderful have been and are being evolved.

> — Charles Darwin,
> *The Origin of Species*

*W*e are here to make sense of sex differences: what they are, how they came to be, why they are important, and what they mean to our everyday lives. All people have sex on the mind, in the brain, and in the body. Whether expressed as a physical experience, an emotional attraction, or gender differences, sex pervades our daily lives. True, people do nonsexual things such as "surf" the Internet, dream up religions and weapons of mass destruction, perform neurosurgery, and write advertising jingles and books about button collecting and computer programming. Sex is not the only thing people think about or do.

But even the Internet, Madison Avenue, and the publishing world, not to mention religion, medicine, and the military, devote prodigious time and space to sex. From the cradle to the grave, whether male or female, heterosexual, homosexual, or bisexual, everyone has sexual experiences of one kind or another. No one is truly asexual: persons with childhood autism still go through puberty, individuals who have been castrated for cultural or medical reasons still have gender identity, and even the avowedly celibate must contend with sexual yearnings.

Very few people, however, are entirely comfortable with sex. Although sexual intercourse itself is something of a learned skill, the relationships leading up to it are even more difficult. Men and women worry about finding a mate, about being attractive to him or her, about having to compete with others seeking the same goal. Courtships go awry; encounters with both opposite-sex and same-sex people are frequently charged with an unaccountable energy, relationships often fall prey to miscommunication and distrust. And so our ability to compete with others, to find and hold mates, and even to function normally is constantly tested, just as all men and women constantly evaluate and reevaluate themselves and one another—not just as people but as *men* and *women*.

Moreover, sex—by which we mean not only the act itself but also sex as it applies to gender-specific behaviors—has an impact on almost every aspect of human existence: intellectual abilities, childrearing, propensity for violence. Even those among us who are not obsessed by sex in its more obvious or prurient forms are deeply and almost constantly aware of sex differences as demonstrated by the simple dichotomy of boys and girls, men and women. Is that surprising? After all, from the moment of birth, when the nurse or doctor utters the portentous "It's a girl" or "It's a boy," each of us is unambiguously assigned to one category or the other. When it comes to gender, there is very little "in between." From the chromosomes in every human cell to the preoccupations of every human life, the world is cleft in two: male and female. "As different," we sometimes say of two things that are clearly distinct, "as night and day."

But just as night and day blend seamlessly into each other, there are some ambiguous cases of sex differences, notably transvestites and transsexuals and even a few hermaphrodites. Because they are so spectacularly unusual, exceptional cases of this sort receive disproportionate attention, whether in movies like *The Crying Game* or Broadway plays such as *Victor/Victoria*. But don't be misled: such gender bending is extremely rare and does not detract from the overwhelming, commonsense fact that sex differences are not only important but real.

Indeed, human beings are probably more aware of the difference between male and female than of any other distinction in the natural or human-made world. We may, on occasion, have a hard time recalling our telephone number or zip code, but anyone capable of communicating can state his or her gender. And although many of us forget—almost instantly—the name of someone we have just met, just as we may also forget the color of his or her eyes, how he or she was dressed, and so forth, we are unlikely to forget whether he or she was a he or a she. Overall, it is difficult to name anything that is more taken for granted than sex differences. Among the more obvious facets of our lives, few things have been analyzed more and yet understood less than what it means to be male or female. To paraphrase Winston Churchill, never have so many been so concerned about something they understand so little.

In our dealings with students, patients, and lecture audiences, we are often asked by women, "Why does my boyfriend want sex more often than I do?" Men ask, "What does my girlfriend mean when she complains that I don't communicate enough with her?" And everyone asks, "Why are men so often violent?" and "Do women think differently from men?" In fact, Judith's office is flooded with patients hoping to resolve gender-related conflicts, people whose sexual lives are fraught with difficulty and frustration. There is the mother who is dismayed by how difficult it is to raise nonsexist children: her daughter prefers to play house, and her son insists on playing soldiers even if he has to line up the forks and spoons to do so. There is the upwardly mobile career woman who fears that her growing assertiveness will turn men off; the recently married man who loves his wife but finds himself attracted to sexy co-workers; the lawyer whose husband dismisses his casual extramarital encounters as "no big deal"; the man who assaults his wife when he discovers that she's been unfaithful. Smart women and men ask why they have selected poor mates and vow to do better next time but haven't a clue as to how to go about it.

The fact that sexual activity is highly desired despite its difficulty and frustration suggests that sex ranks high among the priorities of our species. Consider how much time human beings spend seeking the ideal mate, holding on to tumultuous relationships, and grooming their bodies to appeal to the opposite sex—not to mention the emotional and physical energy invested in courtships that go awry or the anguish suffered as a result of miscommunication with one's partner. And who has not felt the pain of rejection? Yet most of us pick up and carry on, determined to continue the quest for a satisfying sexual relationship.

Our goal in writing this book is not to tell our readers how to have better sex lives, more successful courtships, or sharper communication skills—at least, not in so many words. Rather, we intend to examine fundamental distinctions between males and females and to suggest a unifying biological basis for those differences, thus helping to demystify an important part of our universe and ourselves.

Because sex-specific behaviors are expressed by the simplest lifeforms as well as the most sophisticated, we talk a lot about animals on these pages—not only the birds and the bees but also elephant seals, hyenas, lions, worms, and fishes. We do so in graphic detail because we believe that by closely examining the sexuality of other species we can learn a lot about our own behavior: in particular, why we do the things we do.

In a genuine sense we are all animals, genetically connected to one another through an ancient lineage of species, an intricate web that extends back to the primordial ooze from which life sprang almost 4 billion years ago. Since the birth of the earliest cells, evolution has organized the fabric of life, weaving together nature's warp and woof into a remarkably graceful pattern of history, hardware, and happenstance. Like atomic theory, which provides a unifying basis for understanding chemistry, physics, and biology, evolutionary theory provides a unifying basis for understanding the profusion of life on earth, from paramecia to people. Not only do we physically share our planet with a buzzing, blooming profusion of living things, but we are also genetically linked to our fellow inhabitants, connected to them through an ancient and intricate evolutionary past. Biology looms large in this book because it looms large in all living organisms. As University of Texas psychologist Delbert Thiessen put it, "We do not walk through nature; nature walks through us."

Evidence that biology scripts the human species can also be found by looking to the field of anthropology. Just consider the panorama of human cultural diversity in all its wild and woolly manifestations, from New Guinea highlanders to Lapland reindeer herders to Polynesian fishermen to Afghan pastoralists to Manhattan stockbrokers, and it is easy to build a prima facie case for the organizing and underlying role of biology. Among such an incredible variety of cultures, all with vastly different patterns of social learning, technological development, religious tradition, historical background, and so forth, one common thread emerges: the biological nature of *Homo sapiens* and what it means to be male or female. As we shall see, some societies minimize the difference between the sexes; others—perhaps the majority— exaggerate them. But the differences are never reversed, and thus evidence mounts in favor of a biological common denominator.

Just as ecologists have come to appreciate that all things are connected, evolutionary biologists, too, are starting to recognize previously unsuspected connections between "pure biology," such as eggs and sperm, and the various complex roles and elaborate social "facts" that make up our daily lives. Gametes and gonads, genes and gender all work together to produce sex differences, not only among human beings but among all living things.

Yet the biological pull on our beings is often overlooked; indeed most people are blissfully unaware of the full extent to which biology affects their lives. One recent study of human sperm count attests to the subtle influence of biology. As part of the experiment, ten sexually active couples were physically separated from their partners for various periods of time (during which they remained celibate) and then reunited. When they resumed their sexual relations, the number of sperm in the ejaculate of each man increased in relation to the length of time since previous intercourse. Such findings are not in themselves surprising: sexual abstention is known to raise sperm levels on resumption of intercourse. But noteworthy is the fact that no such increase occurred in ejaculate obtained through masturbation after identical periods of abstinence.

Apparently, there exists a factor—hidden and previously unrecognized—that results in greater sperm production or transfer during sexual intercourse than when fertilization is not a possibility. Although the mechanism remains unknown, the phenomenon makes evolutionary sense. Why waste sperm (which take energy to produce) when no offspring are in the offing? We cite the foregoing example not because it is overwhelmingly important in itself but because it helps reveal how even *Homo sapiens*—smartest of all animals—can be influenced by evolutionary pressures without having the slightest idea that anything of the sort is going on.

For those inclined to denounce a biological approach to understanding male–female differences as sexist, we hope to reveal that if anything, sexism comes from culture, not from biology per se. Social learning and cultural traditions can magnify or suppress sex differences in human beings by rewarding certain behaviors and condemning others, as well as by providing models and expectations for ways in which boys and girls ought to behave. But they no more create those basic differences, any more than they create the basic biology of maleness and femaleness. Society is responsible for establishing sexist social roles and expectations (what are traditionally called *gender differences*). But it is evolution that makes for *sex differences*, the basic, organic, genetically inspired biological distinctions between women and men.

When we lecture on male–female differences in sexual style and motivation, we are often confronted by members of the audience who triumphantly maintain that we must be wrong. Why? Because they know a man, for example, who is altogether nonaggressive or sexually reticent, or a woman with a killer instinct or whose libido has her chasing every male in town. The generalizations in this book are just that: generalizations. A generalization, by definition, applies to the majority within a population, allowing plenty of room for individual exceptions. It is perfectly true, for example, that men generally weigh more than women. This does not mean that there aren't some small men and large women; it simply means that at the level of the population, the weight of a randomly chosen man will usually exceed that of a randomly chosen woman.

As we make generalizations throughout this book—about sexual inclinations, parental tendencies, aggressiveness, and so forth—please keep this simple body-weight example in mind. Thus, when we point out that men are more likely than women to become sexually aroused by simple visual images, we are not claiming this to be true without exception. We are not proven wrong simply because some men are indifferent to *Playboy* centerfolds and some women are turned on by posters of Antonio Banderas. Rather, we are talking about general and widespread tendencies, no different in principle from other male–

female distinctions that are universally accepted, such as body weight, vocal range, number of hair follicles, and the like.

We do not intend to be judgmental and have tried to steer clear of declaring certain behaviors good or bad. If, again, men are more aroused than women by nude photographs, does this mean they are more barbaric or primitive, unable to distinguish image from reality, or less loving? Or that women are sexually repressed, inhibited, or otherwise defective? Neither. It simply suggests that natural selection has created different behavioral inclinations.

As with so many difficult questions concerning human beings, it may be impossible to prove absolutely that biology exerts a powerful pull on our personas. But we think this book makes a strong case for that assertion, one that will help our readers to make sense of sex and sex differences and, in the process, to better understand their lives. Furthermore, we hope that the insights we offer into the complexities of human behavior and the evolutionary roots of maleness and femaleness will help increase readers' sensitivity to their fellow human beings. We aim to assist the reader in acquiring self-knowledge and also to demystify the opposite sex—those crucial others with whom each of us shares so much and yet who are often so infuriatingly and fascinatingly different.

CHAPTER 2

Biological Roots

DESCENDED FROM MONKEYS? My dear, let us hope it isn't true! But if it is true, let us hope that it doesn't become widely known!

> — wife of the Bishop of Worcester,
> 1860, on being told of the
> scandalous work of Charles Darwin

"Biological differences between men and women?" some of our readers might well exclaim today. "Let us hope it isn't true!" But there is every reason to think that it is true and, furthermore, every reason to suspect that evolution has had a strong hand in producing those differences.

The reader eager to harvest an instant armload of glib, quick-and-easy generalizations about men and women will not find them in this chapter. We do talk a lot about human beings, but we also present a lot of theory as well as factual information about other species. It turns out that the myriad

ways in which human beings go about their sexual lives—from courtship to mating to parenting to interacting with those of the same sex—
very much mirror what thousands of scientists working for many years
have seen in other species.

As an evolutionary biologist and a physician, we are not alone in
seeking to enhance human self-understanding by paying careful attention to other living things. Neurophysiologists learn a great deal about
human brain function, for example, by studying the oversized nerve
cells of the giant squid. Geneticists hone their understanding of DNA
by tracking the genetic vagaries of the lowly intestinal bacterium *E. coli*.
Hardly any living things are as far removed behaviorally from human
beings as giant squid or intestinal bacteria, and no reputable scientist
would argue that we should extrapolate directly from either to *Homo
sapiens*. What we can do, however—and what ethologists, sociobiolo-
gists, behavioral ecologists, evolutionary psychologists, call them what
you will, have been doing for several decades now—is to look carefully
at bacteria, and squid, flycatchers, and elephant seals, as well as anything else that catches the eye and stimulates the mind, in an attempt
to discern some of the underlying rules that govern life. Just as biologists have studied other living things to reveal how neurons communicate and how DNA replicates, in this book we shall examine other creatures to see how evolution works to make males and females.

In this chapter, we also grapple with issues most people take for
granted, delving into why sex evolved in the first place and why the sexual world is divided into two instead of three, four, or any other number of sexes. Most important, we look at how these events set the stage
for male–female differences, differences that are frequently distilled
into such quips as "Oh, that's just like a woman" or "It's a guy thing."
The rest of the book, in which we consider specific arenas of male–
female differences—sex, violence, parenting, childhood, and so forth—
builds on the ideas presented here.

The Power of Evolution

Human beings are far from biological automatons. We have free will.
We do want what we want, within societal constraints. And yet at the
same time, we are influenced by many factors, including our biological
heritage. At least in part, we behave as we do—not consciously, but on

a deeper, more instinctual level—because some sex-specific behaviors have been rewarded by the forces of evolution. In other words, men tend to behave in certain ways and women in others because in the distant past, their sex-specific behaviors gave them an evolutionary advantage. The process by which distinctive male–female behaviors arose is an important part of evolution by natural selection.

Simply put, evolution favors any genetically determined trait—whether behavioral, structural, physiological, or psychological—that leads to more offspring, who in turn pass on those traits to the next generation. For example, a man who is especially attractive to women is more likely to reproduce than one who is not. Thus, more of his genes are likely to be projected into the future and to become more numerous in succeeding generations. It's as simple as that. There is no willful agency or grand plan; natural selection is simply the result of fortune, in which the winners are those individuals or genes that are most reproductively successful. (By way of comparison, artificial selection occurs when humans breed plants or animals deliberately to promote various traits, whether the huge size and docile disposition of the St. Bernard or the durability of the standard grocery store tomato.) Either way, the fact that some individuals and their genes do better than others translates into something of importance.

Individuals who produce lots of offspring, who in turn produce lots of offspring, and so on from generation to generation, are said to be more biologically or reproductively "fit" than their competitors. Biological fitness—not to be confused with physical fitness—is measured in terms of reproduction and generations. One way to think about the concept is to compare two couples. A man and woman who produce four children but lose three of them during childhood end up having lower biological fitness than a couple with two children who both survive and produce children of their own. This would be true even if the first couple were Olympic athletes or lived to be 100, or if the second couple died young. All that counts in biological fitness is the number of genetic copies—generally, offspring—that survive over many generations. The implications for human sexual behavior are immense.

In practical terms, evolution should favor—to a degree—individuals with strong sex drives. People who had sexual intercourse the most frequently and successfully would tend to have the most offspring (especially in the days before birth control). In turn, the offspring would probably inherit a strong sex drive and would themselves have more offspring, and so on, until most people in the population had robust sex drives. By contrast, those with lagging libidos would probably be comparably sluggish in passing genes on to subsequent generations and would therefore be less represented in each passing generation.

Indeed, sex seems to be a central force in the lives of most animals. Teenage boys, on average, are said to think about sex once every three minutes; aging women may spend enormous amounts of money and time trying to hold on to their younger, sexier selves; and impotent men often go to great lengths to restore their virility. Why all the emphasis on our sexual beings? The question is particularly relevant in light of the enormous price tag affixed to sex—for as we shall see, sex is expensive, and not just in dollars and cents.

Why Sex?

Sex remains one of the great mysteries of evolution; even today, biologists do not know precisely why it evolved. "Is sex necessary?" inquired James Thurber and E. B. White in a delightful book by that title. The answer—strictly speaking—is a definite no.

A number of species have dispensed with sex altogether. Many single-celled organisms—bacteria, amoebas, and so on—procreate simply by splitting in two. Some plants, such as strawberries, send out runners. Other species have lifestyles that rival science fiction. For example, the hydra, a sedentary relative of the jellyfish, reproduces by budding: a miniature hydra sprouts forth and then breaks off from its parent. (Imagine an analogous situation in human beings, with women giving birth by growing their babies as external appendages!) Still other animals, including some insects and even a vertebrate or two reproduce by a process called parthenogenesis, in which eggs develop without being fertilized. Among certain species of whip-tailed lizard, for example, females lay fertile eggs without the benefit of sex or sperm. Interestingly, female whip-tails sometimes take turns performing the male role in courtship, which encourages the partner to produce more eggs. And yet, in all these cases of so-called asexual reproduction, there is no mingling of the flesh, no exchange of bodily fluids, no mixing of genes; in short, no sex.

For humans and most other animals, however, sex is a requirement of life; without sex, our species would come to a crashing halt. Of course, sex in the colloquial sense of having intercourse isn't necessary for making babies. Women can become pregnant via artificial insemination or in vitro fertilization, but biologically each of these women is "having sex" nonetheless. Indeed, we are programmed by our evolutionary past to seek sex, to revel in its sensory aspects, and to engage in it frequently, certainly more than is needed to reproduce. Although some people have sex to make a baby or because of pressure from their partner to do so, most of us have sex because it feels good. Why?

Physical and emotional intimacy, touching and being touched, excitation, ejaculation, orgasm: all are rewarding in themselves. But this explanation provides only a partial answer. Like eating in response to hunger, the drive and pleasures of sex are "proximate mechanisms" providing us with an immediate reward and thereby ensuring that we do those things that lead to an ultimate evolutionary payoff. But why does sex enjoy such favored status in our lives? What gives ultimate direction to our erotic inclinations?

Ultimately, the answer has to do with diversity. Sex spins forth diversity. Without sex, individuals would be exact replicas of the parent who produced them: rather than carrying half the genes of one parent and half of the other, our descendants would have 100 percent of our genes. Put another way, an individual who reproduced sexually would have to bear twice as many children to keep up with someone who cloned herself asexually. Sex therefore automatically carries with it a heavy genetic cost: it dilutes an individual's genetic legacy by 50 percent.

At first glance, giving rise to offspring identical to oneself might sound rather attractive—at least to those of us who are fairly happy with our genetic lot in life. All the recent hubbub over Dolly, Ian Wilmut's cloned sheep, is a good indicator of how deeply fascinated humans are with the not-so-surreal possibility of someday creating a genetically identical self. Short of discovering the fountain of eternal youth, what could be closer to achieving immortality, either for oneself or for a loved one? Apart from narcissism, in a world governed by natural selection, wouldn't self-replication be favored because it allows 100 percent of one's genes to be passed on to one's offspring, rather than only half?

Not really. The key is that life is never static. The environment in which we live is ever changing, and asexual reproduction offers little protection in such a continuously shifting world. Picture an individual superbly adapted to its natural setting—a mottled green-and-yellow frog, perhaps—whose exquisite pattern of coloration provides perfect camouflage amid the mixed rushes and cattails of its watery habitat. But if the climate shifts—as it inevitably will—and the rushes give way to grass, the frog's identical descendants will find themselves exposed to the watchful eyes of predators. And so a population of identical frogs might all be eaten, whereas among a crowd of genetically different frogs, the chances are that one or more would have enough green to be camouflaged among the grass and would prosper.

In human terms, the value of genetic diversity becomes apparent whenever a few people squeak by in a deadly epidemic or a natural catastrophe. Some survival may be circumstantial, but much of it can be attributed to genetic variability. Indeed, some biologists think that sex evolved as a way of producing moving (that is, constantly changing) targets, which are more difficult for parasites and disease organisms to hit.

Genetic variation is analogous to a lottery. Rather than buy ten lottery tickets with the same number, it behooves the prudent gambler to purchase ten tickets each with a different number, thus increasing tenfold his or her chances of winning. In the same way, it seems foolish to put all one's reproductive eggs in a single basket. By investing in genetically different offspring, one is more likely to end up with at least a few winners in the lottery of life.

Regardless of its evolutionary rationale and whether it is necessary or unnecessary, adaptive or maladaptive, sex remains central in the lives of most human beings. The fact that it looms so large suggests that its biological roots run deep. But, as we shall see, teasing apart those roots and coming to a better understanding of the toll sex takes—especially on women—can help us better understand our own lives: why, for example, women are more hesitant than men about jumping into bed. Overall, without sex, life might be less fun, but it would also be safer and more predictable, at least in the short term.

The Cost of Sex

Leaving aside the risks of pregnancy and childbirth, sexual reproduction by no means guarantees a happy outcome for those who engage in it. Rather, it is fraught with problems, posing significant risks to both males and females.

To begin with, sexual reproduction requires a mate. Most humans will acknowledge that the search, at least the search for a permanent mate, can be frustrating and prolonged, riddled with false leads and inappropriate choices. But the process may be especially difficult for an endangered species whose few remaining members are widely scattered. In such cases, finding a mate—any mate—becomes downright desperate. The situation reminds us of our favorite imaginary creature, Arnold, the sad Long-Necked Preposterous, from Shel Silverstein's hilarious children's book *Don't Bump the Glump!* Poor Arnold spends his days "looking around for a female Long-Necked Preposterous. But there aren't any." Individuals who could reproduce on their own not only would avoid Arnold's dilemma but also would have a leg up in the race against extinction.

In addition, sexual reproduction operates much like a game of roulette. When the wheel is spun—and for females, the cost of spinning is especially high—there is no guaranteed outcome. Common lore has it that the famous and beautiful dancer Isadora Duncan once propositioned crusty old George Bernard Shaw, suggesting that with her looks and his brains they would have a most splendid child. Shaw sagely rejected Duncan, pointing out that it was equally possible that their child would have his looks and her brains. In fact, it is even more probable that Duncan and Shaw would have produced a rather average child, neither as verbally brilliant as Shaw nor as physically talented as Duncan. This so-called regression to the mean might well have disappointed both parents, each of whom would have hoped for a child possessing their better traits.

Not only does sexual reproduction offer no guarantee as to which genes will end up in which child, but it also chooses genes randomly from each parent and then rearranges them to yield each new individual. As all parents know, there is nothing predictable about having children; a man hoping for a son who looks just like him can just as easily end up with a son who resembles his maternal grandfather and has the personality of his mother.

Reproductive roulette can also create genetic nightmares. Consider the anguish of parents whose children are diagnosed with genetic disorders such as cystic fibrosis and sickle-cell anemia. In a way, modern technology adds to their misery by making it possible to pinpoint the source of a biologically inherited disorder and thus place genetic blame on the carrier of the disease. In fact, parents of children who have inherited a serious genetic illness from the other parent often express simmering resentment toward the spouse.

Judith has seen these emotions poignantly expressed by her patient Christina, a brilliant young medical scientist. Christina's son, Chris, has attention deficit hyperactivity disorder (ADHD). She believes—correctly, it appears—that Chris inherited this illness from his father, who has struggled with it all his life. Jerry, Christina's husband, is a gifted musician who didn't do well in school because of poor work and study habits. He is messy to a fault, impatient, and impulsive. Even Jerry's band barely puts up with his lateness and unpredictability, but his compositions are so vital and brilliant that his fellow musicians reluctantly tolerate his eccentricity. Christina loves Jerry for his energy and devotion but pays a big price for this love. Not only does she work full-time to bring in a steady income; she is also the only person in the household who organizes closets, does the cooking, and handles the family finances. Jerry daydreams at his synthesizer while Christina makes beds and unpacks groceries.

Chris is an extraordinarily bright boy, but because of his illness he needs an enormous amount of time and attention. Christina faces a dilemma: how much energy, how many hours in the day should she invest in herself and her own career and how many in her son? Should she try for a second child, hoping that the experience will be more gratifying than the first but knowing that she runs the risk of having two children with ADHD? As she considers her options, she confronts her anger that she mated with someone with a genetic disorder and as a result, the child she loves is impaired and her own life is heavily burdened.

Another sobering fact of sex is that women physically pay a high price, far higher than men, if only because during intercourse the woman is alone with someone who is usually bigger, as well as temporarily crazier, than she is. The human female may entice, solicit, and even initiate sexual intercourse, but during the act she submits to having her body penetrated by someone who is generally stronger and also more fervent. (We haven't yet mentioned the heaviest and most portentous costs of sexual intercourse, which fall entirely on women: pregnancy and childbirth.)

Among humans, rough sex can vary from petty inconsiderateness and thoughtlessness to rape and overtly sadistic practices; in any event, the victim is almost inevitably on the receiving end of the phallus. Men are rarely injured by intercourse. Other than an occasional myocardial infarction during orgasm or minor muscular twitches, intercourse for men is, on the whole, far more pleasure than pain.

Even in high-quality, emotionally intimate relationships, women are prone to urinary tract infections and what is clinically called dyspareunia, or pain during intercourse. Few heterosexually active women do not at least occasionally submit to undesired sex, putting up with unwanted exertion or even discomfort or physical pain rather than disappointing or negotiating with their partners.

Judith recalls the anguish in a patient's voice as she described what amounted to date rape. The young woman had willingly gone for an evening stroll along a beach with an attractive man; had eagerly cuddled with him, sharing a bottle of wine as they watched the moon rise over the ocean. What started as consensual sex then turned violent. The next day he apologized profusely and asked her to give him another chance. Not only did she suffer physically from the encounter, but she was also mentally bruised and emotionally confused. Had it been her fault? She had knowingly flirted with the man and had happily gone with him to a secluded part of the beach, let herself get drunk, and initially agreed to have sex. Or was he to blame? He had chosen the secluded location, plied her with wine, and then ignored her cries to stop when he became violent. Could he truly be sorry for what he claimed was a one-time loss of control and never do anything like it again? Should she give him a second chance, as he was begging her to do?

Conversely, in some animal species the sex act may be riskiest for the male, although such cases are relatively rare. The most dramatic examples involve certain spiders and preying mantises in which the female eats her mate during copulation. Among some fireflies, females mimic the flash patterns of other firefly species; when a male lands next to one of these femme fatales expecting to mate, she eats him. And rough, if not lethal, sex is characteristic of many vertebrates, including members of the cat and weasel families. Horses, too, often bite or kick one another as part of mating.

In addition, sex can lead to increased mortality by exposing a mating pair to predators. All the calling, singing, roaring, prancing, posturing, dancing, billing, cooing, leaping, and fluttering that so occupies lovers among insects, fish, reptiles, birds, and mammals sends a strong signal to their enemies. At the same time courting males and females are homing in on each other, predators are homing in on them. Some predators specifically target individuals that are engaged in sexual behavior. Zoologist Michael Ryan of the University of Texas has described a species of Central American bat, for example, that locates frogs by their mating calls. Male frogs are thus faced with a cruel dilemma: croak loudly and take the chance of being eaten or keep quiet and remain celibate.

If violence and genetic liabilities were not enough, sex carries other costs, including sexually transmitted diseases, which affect both women and men. Syphilis and gonorrhea are ancient scourges; the former was widely believed to have been introduced into Europe from the New World by Columbus's returning crew. Genital herpes made a strong showing during the 1970s. And, of course, AIDS has emerged, in many ways, as *the* disease of the 1980s and 1990s, carrying with it an enormous burden of suffering and loss. Human beings are by no means the only animals to suffer from venereal diseases: for example, monkeys suffer from a form of HIV, and horses are subject to a variety of sexually transmitted diseases. All in all, the term *safe sex* exists only because sex is inherently unsafe.

Some people claim there is a lesson to be learned from such epidemics: monogamy. Others say that abstinence is the only true safeguard. Our point is more general: if human beings were majestically aloof, able to reproduce in solitary splendor without sex as do strawberries and dandelions, they would be spared such moralizing and—more important—such mortal dangers. The brutal, biological reality is that sex requires intimate contact with others, and as sexually transmitted diseases make tragically clear, such contact is bound to be risky. The fact that sexual intercourse is so prevalent despite such drawbacks simply underscores its biological importance.

A World Divided in Two

Why is the world so neatly divided into males and females? Nearly all animals, as well as most plants, have two recognizable sexes: male and female. Every once in a while, one encounters a person whose gender is difficult to determine by casual examination. Most of the time, however, there is little doubt about an individual's sexual identity. Clothing and hair styles are important clues, as are physical characteristics. Broad shoulders and narrow hips suggest a man; prominent breasts and a narrow waist suggest a woman. A naked person, of course, leaves no room for doubt: a man's penis is distinctly different from a woman's labia.

But what about birds, such as robins or warblers, in which both sexes have an identical genital opening, the cloaca? Even in these cases, ornithologists can distinguish male from female with absolute certainty. To be sure, these animals have other external distinctions, such as the generally brighter color of the males, but such physical traits do not define maleness or femaleness; rather, they are consequences of the male–female difference. Ultimately, it is the type of gamete—egg or sperm—an individual produces, rather than penis or vagina, breast or beard, color or costume, that determines the difference between maleness and femaleness. Herein lies the crucial difference between bulls and cows, stallions and mares, men and women. Eggs, which are large and relatively few in number, are produced only by females, whereas sperm, which are small and abundant, are produced only by males.

Still, why has evolution split the sexual world into two parts instead of three or even thirty-three? Alternatively, why not dispense with sex differences altogether and simply have one sex? Why bother with men and women, males and females, eggs and sperm? After all, sexual recombination—the coming together of genes from two different individuals—requires only a shared willingness by the parents to bequeath some of their genetic make up to the next generation. Two same-sized cells from different parents should do the trick. There is no obvious biological reason why a species couldn't consist entirely of unisex members, with everyone producing medium-sized gametes. In fact, the sexual possibilities would increase in such a system, since everyone could, theoretically, mate with everyone else instead of the current arrangement, in which half the members of each species are out of bounds.

Yet evolution has clearly favored the existence of two sexes, with one producing large gametes and the other, small ones. The size difference between eggs and sperm can be staggering, in fact, with some animals making eggs 100,000 times bigger than their sperm. Even in humans, eggs weigh 85,000 times as much as sperm. When it comes to sheer bulk, the ostrich egg, which weighs as much as three pounds, takes the prize, though some birds produce eggs that constitute more than 25 percent of their body weight. Imagine a human female producing a thirty-pound egg once a month!

By contrast, sperm are negligible in size—on average, one ten-trillionth of a gram—each little more than a tiny package of DNA outfitted with a long, lashing tail. If the head of a rooster's sperm could be blown up to the size of a hen's egg, its tail would be about a yard long; the egg, comparably enlarged, would then be nearly a mile across! The difference is less extreme, but still considerable, in the case of mammals.

How did evolution produce such drastically different gametes? British biologist Geoffrey Parker created a compelling computer model that gives a pretty good answer. Parker started with a hypothetical group of sexually reproducing individuals, but with no identifiable males and females. Because some variation occurs within all sexual populations, Parker's model assumed that some gametes would be larger and some smaller, although the differences among them would be slight. The larger ones would have more yolk, giving embryos a better start following fertilization, but they would also be somewhat less mobile and therefore less capable of seeking out other gametes.

Meanwhile, the smaller gametes, containing fewer nutrients, would be more mobile and therefore more likely to bump into other gametes. Selection would strongly favor those that matched up with larger partners, a fusion likely to result in nutrient-rich embryos. Over time, the smaller, poorer, but peppier gametes would specialize in seeking out the larger, richer, slower ones and would compete with other small gametes trying to do the same. Because the small, active gametes are cheap to produce, they could be turned out in large numbers, providing plenty to go around. After many generations, as the larger gametes became yet larger and more richly endowed and the smaller gametes became even smaller and faster, the distinctions between them would become absolute. And so from the birth of eggs and sperm came a fateful, portentous step in the history of life: the evolution of specialist gamete makers, to which we give the familiar names *male* and *female*.

Males then evolved bodies and behavior appropriate for sperm makers; females evolved bodies and behavioral repertoires suited to their specialization as egg makers. The technical term for this is *sexual dimor-*

phism (from the Greek *dimorphos*, "two shapes"), and as we shall see, it applies to behavior and bodies no less than to gamete making.

What Makes Males Tick

Overall, males have little to do with the actual business of reproduction, beyond producing sperm packaged in seminal fluid. In contrast, all female mammals invest enormous resources in their offspring after fertilization occurs. They not only build a placenta and nourish the developing fetus but also nurse their young after birth. Admittedly, many women today do not breast-feed their children, but most human tendencies were established thousands of generations ago, long before modern technology and current cultural whims.

The notion that differences in parental investment, with females producing costly, nutrient-rich eggs and males producing cheap, near-naked sperm, account for behavioral differences between males and females was first articulated twenty years ago by Robert Trivers, a young evolutionary biologist then at Harvard University. He postulated that sexual competition is a replay of fertilization itself, in which numerous males, like small, hyperactive sperm, compete among themselves for access to females. Success crowns those who are pushy enough to outcompete their rivals yet have enough wanderlust to keep moving, searching for new conquests. But unlike fertilization, in which (as far as we know) the egg passively receives suitors, females are usually too invested in their potential offspring to mate with the first male that happens along.

Roving Inseminators. As Trivers described it, the sex—in most cases, the female—that invests more in offspring becomes a limited resource, something for which the sex that invests less—usually, the male—must compete. Thus, in evolutionary terms, it behooves males, which can produce millions of sperm with relatively little effort, to mate with as many females as possible. But females, which produce far fewer offspring, benefit most by being selective—that is, by choosing mates that are healthy, strong, smart, wealthy, and so on.

A useful analogy can be found in the financial world, in which males can be compared to stockbrokers and females to wealthy investors. An investor's fat wallet would doubtless attract a number of stockbrokers, each eager for access to a profitable portfolio. But the investor is well advised to be choosy and to think carefully before committing her wealth exclusively to a single broker. As a consequence, investors and brokers behave very differently, the latter competing among themselves for access to the former and trying, more or less indiscriminately, to line up as many clients as possible. In this analogy, a good investor would see through the hoopla and false advertising and choose only an honest and savvy broker.

Males evolved, in a sense, by parasitizing and taking advantage of the metabolic investment provided by females. In strictly biological terms, they are little more than roving inseminators, well skilled in the bluffing and blustering, pushing and shoving, shouting and showboating that gives them access to females. Put in human terms, men, whether they admit it or not, enjoy sexual conquests and are highly competitive with one another, at least subconsciously, as they attempt to overcome, persuade, and entice women.

Not surprisingly, males are also easily aroused. Who wants to miss out on sex by being too slow on the draw? Among frogs, for example, males are so notoriously eager to dispense their sperm during mating season that many will clasp almost anything—a softball, a soda bottle, another male—that doesn't clasp them first. (Some even have a special croak that says, "Let go. I'm a male, you dummy!") In contrast, females are careful investors, concerned more with quality than with quantity of offspring, and so they choose their mates carefully. The result: males end up competing—sometimes furiously—with other males for access to a limited number of females.

These different reproductive strategies—and the theory of parental investment that enables us to understand them—go a long way toward explaining male–female differences. Time and again, patterns of social and sexual behavior among animals have been found to correspond to the patterns predicted on the basis of differences in parental investment.

Harem Masters. Consider elephant seals, well known for their gargantuan males, which at 6,000 pounds weigh four times as much as the average adult female. As is typical of species with striking sexual dimorphism, elephant seals form harems, in which a cluster of females mates exclusively with a dominant male. Parental investment on the part of the female is extremely high (an adult female weighing about 1,500 pounds typically gives birth to a 110-pound pup, which may then gain another 220 pounds from its mother's milk), and she is nearly always mated by a harem master.

As might be expected, the male elephant seal invests virtually none of his immense bulk in offspring. Instead, he throws his weight around in other ways, trying to inseminate as many females as possible. A successful bull elephant seal may have forty mates, each of which will probably bear one pup per season. But because the sex ratio is one to one, for every harem master there will be thirty-nine disappointed bachelors. Although a few males will be immensely successful, producing some forty pups per season, most will be reproductive failures, living their lives on the sidelines, with no mating prospects. In short, males operate within a system that is inherently unequal, divided between reproductive haves and have-nots. Because the stakes are so high, competition among males during mating season is intensely physical, often expressing itself in violent clashes that result in serious, though rarely fatal, head and chest wounds.

From a strictly biological point of view, human males are much like bull elephant seals and many other male mammals: they are larger and more aggressive than females, become sexually mature later, and have higher mortality rates. Such traits exist because of the evolutionary advantage they confer on their owners: larger size and heightened aggressiveness are likely to win more mates, by brute force, if nothing else, and thus to produce larger numbers of surviving children. And delayed maturation enables males to avoid violent competition when they are too small, too inexperienced, too weak—that is, too young— to prevail. In humans, this particular sex difference is immediately apparent in any seventh-, eighth-, or ninth-grade classroom, where the more mature girls tower over boys who suddenly—if temporarily— seem several years younger than their classmates.

Populating the Planet. Like other male mammals, each man is theoretically capable of producing a large number of offspring while being obliged to contribute very little in the way of parental investment. Consider the fact that in a single ejaculation a man releases 200 to 600 million "pollywogs,"—enough, in theory, to fertilize every woman in the Western Hemisphere. (Although this may seem like a lot, note that horses produce 4 billion to 9 billion sperm per ejaculation and pigs, as many as 20 billion.)

One way to appreciate the biological relevance of such productivity is to imagine that every man on earth except one is suddenly killed. Although mighty exertions on the part of the surviving man would be necessary, as would artificial insemination techniques combined with appropriate dilution of his sperm, a large number of women probably could be made pregnant, and fairly quickly. In fact, assuming he was relatively healthy and not too old, the surviving Adam could almost certainly be the progenitor of thousands of children, probably tens of thousands, all within his lifetime. Now, reverse the scenario and imagine that every woman on earth except one is suddenly killed. How many children could the hypothetical Eve bear? How long would it take for the human species to regain its previous numbers?

Lest this seem the stuff of male fantasy, real life has its own astounding examples. King Sobhuza II of Swaziland had more than 100 wives, about 600 children, and thousands of grandchildren; it has been estimated that one-fifth of all Swazis are his descendants. In contrast, the known record number of children born to one woman is 69—a remarkable figure that includes multiple sets of twins and triplets. Our point is that males have an almost unrestricted ability to reproduce, whereas females face straightforward biological limits when it comes to babymaking.

This difference has important consequences, as reflected in the following ditty:

> Higamous, hogamous, woman monogamous.
> Hogamous, higamous, man is polygamous.

This little bit of poetic profundity has been attributed to psychologist William James, supposedly with an assist from opium, though James may have had something of an intuitive whiff from evolutionary biology as well. As we have seen, men are reminiscent of bull elephant seals: some have more wives and children than is "their share," whereas others have fewer. Women, however, are very likely to have one husband and at least some children, though not a very large number of them.

The tendency for men to form harems and for women to join them is also reflected in marital arrangements typical of our species. Of 849 societies examined in anthropologist George P. Murdock's classic *Ethno-graphic Atlas*, 709 were polygynous (a more precise term than polyga-mous, for it refers specifically to one man having many wives), 136 were monogamous, and only 4 were polyandrous (one woman with many husbands). Time and again, anthropologists have come up with similar results, concluding that before the spread of Judeo-Christian doctrine, polygyny was the preferred marital system for more than 80 percent of human societies.

When biologists see certain characteristics in other species—larger male size, greater aggressiveness, delayed maturation, and shorter life span—they know the males must be competing for access to females. Combined with the overwhelming cross-cultural data on *Homo sapiens*, we can safely conclude that in our history, human beings were polygy-nous. Even so, as with elephant seals, few men had what it took to become harem masters, so most had to settle for either monogamy or celibacy. But the end result was the same: men competed among them-selves—as they do today—for women. When it comes to the biologi-cal whisperings within, humans still tend toward polygyny. And with polygyny comes competition.

Fighting for Females. How much competition at the physiological level takes place in other species remains largely unknown. Biologists do know, however, that in most species males compete intensely for the sexual attention of females. In fact, as long ago as 1871, Charles Dar-win introduced the concept of "sexual selection" to explain why males tend to be relatively large, conspicuous in color and behavior, and endowed with intimidating weapons (tusks, fangs, claws, antlers) and a willingness to employ them in their efforts to gain access to females.

Darwin's idea—upheld today—is that sexual selection, which is a subset of natural selection, favors any attribute that enhances an indi-vidual's chances of reproduction. Traits such as a peacock's tail or a buck's antlers will be favored by sexual selection if the mating advan-tage they confer outweighs their survival disadvantage. The peacock's tail may become tangled in shrubbery and may draw the attention of predators, but if it charms the peahen—and it does—then on balance it will be favored by evolution. The same goes for weaponry in the form of tusks and other hefty adornments. Imagine the energetic costs to a male moose, which carries upward of 50 pounds of antlers on his head, or to a bull elephant, whose two tusks may have a combined weight of 200 pounds!

But tusks, antlers, horns, and the like enable their bearers to outbattle lesser-adorned competitors, and with victory come more opportunities for reproduction. And large-tuskers beget more large-tuskers, and so on. Thus, over time, sexual selection has led to a remarkable preponderance of extraneous accoutrements among males, including wattles, ruffs, collars, crests, iridescent patches of feathers, horny excrescences, bony shields, elaborate antlers, raspy noise makers, perfumed plumes, and an equally outlandish behavioral repertoire designed to defeat other males and attract females.

Such adornments would have little effect on reproduction, however, if their male bearers lacked behaviors to match or if they failed to evoke a response from females. Thus, sexual selection has two behavioral components: first, competition among members of the same sex for access to mates, and second the exercise of choice when members of the more desired sex choose mates from a competitive pool of suitors. Because of the male–female differences in parental investment, competition becomes predominantly a male activity and choice becomes a female prerogative, with fussy egg makers discriminating among pushy competing sperm makers.

Competition among males may be fierce, notably so in polygynous species such as elk, moose, elephant seals, and gorillas, for which being male means ending up either a harem master or an evolutionary failure. Not surprisingly, with such intense pressure, selection favors males that grow up to be large, tough, and well armed: unpleasant bullies, as befits a winner-take-all lifestyle.

Battle of the Sperm. But competition among males expresses itself in more devious and indirect ways as well. In the black-winged damselfly, a common streamside insect of the eastern United States, females mate with more than one male. Each male black-winged damselfly sports a specialized penis outfitted with lateral horns and spines, not unlike a scrub brush. According to Jonathan Waage of Brown University, a copulating male uses his penis to clean out 90 to 100 percent of his predecessor's sperm before depositing his own. In other insect species, part of the male's body literally breaks off following copulation, forming an organic "chastity belt" to prevent the female from mating with anyone else. Some male sharks give their sexual partners a precoital douche, courtesy of their remarkable double-barreled penis. One barrel contains a specialized tube that acts as a high-pressure saltwater hose, sluicing away any sperm deposited by a sexual rival; the other barrel transports sperm into the female.

Perhaps the weirdest case of male–male competition plays itself out among certain parasitic worms that live in the untidy confines of rats' intestines. Here, males mate with females and then seal their union with a special substance squirted from their so-called cement gland, which keeps their sperm from leaking out while preventing other males from getting in. In a devilish twist, males also inject their cement into the genital openings of other males, thus rendering them impotent. In the otherwise sober scientific paper that first reported it, this strategy was described as homosexual rape. But it is clearly not a case of mistaken identity because the aggressor does not inject sperm, as he would when consummating a relationship with a female. Rather, it is strategic. Since only a limited number of females share each not-very-commodious rat gut, and other males want access to these females, a male's chances of reproducing are enhanced each time he knocks another male out of commission. In a fantasy worm world, a male would succeed in plugging up all his rivals and be left alone in a dazzling universe of available females.

Some biologists speculate that a man's long penis might help introduce sperm deep into the vaginal tract, an adaptation to human bipedalism and upright posture, which has caused the female's pelvis to be somewhat tilted from the angle found in most other primates. But no one has proved this to be so. If the reason for the size of a man's penis remains obscure, the same cannot be said for testicles. Animals that have small testes tend to muscle their competitors out of the way, whereas those with large testes cannot dominate a single female but must share her with other males.

Taken as a percentage of total body weight, human testicles weigh twice as much as those of orangutans and fully five times as much as those of gorillas. But in this dubious realm of anatomic competition, the overwhelming champion is the chimpanzee, whose testicles are more than three times larger than those of men and fifteen times larger than those of gorillas. Gorillas have no need for large testicles: a dominant silverback male gorilla that drives away potential competitors by virtue of his physical strength is pretty much guaranteed sole access to the females in his troop. Therefore, his testicles need only be large enough to make sufficient sperm to fertilize the females he has accumulated.

Chimpanzees, however, live in multimale groups in which many males often copulate with a single estrous female. Compared with gorillas, male chimpanzees compete less with their bodies and more with their testes, leaving much of the competition to their sperm. As a result, a female chimpanzee's vaginal tract can be compared to a miniature Roman Coliseum within which tiny spermatic gladiators vie to outfox, outrun, or simply overwhelm the opponents' sperm. Under these conditions, he that produces the most sperm has a distinct evolutionary advantage.

Whales show a similar pattern. The world's largest animal, the 150-ton blue whale, has relatively small testes: together they weigh only about 200 pounds In contrast, the right whale, which is about half the size of the blue whale, has testes that may exceed nine feet and weigh about 1,000 pounds! Like chimpanzees, right whales are relatively promiscuous: many males copulate with the same female, leaving their sperm to battle it out internally.

In most cases, noncompetitiveness among males results in reproductive failure. In the famous children's story of Ferdinand the Bull, this physically impressive creature preferred smelling flowers to fighting other bulls. But few Ferdinands occur in real life because such individuals would be less likely to promote themselves or, more to the point, project their peace-loving, flower-sniffing genes into the future. Regardless of his testes, if Ferdinand has no heart for male–male competition, his sperm—and thus his preference for flowers over fighting—will probably be replaced by that of his less docile rivals. And so it is for humans: how many weak, cowardly, or ineffectual men are sexually attractive to women?

Role Reversals

As with almost every facet of the living world, there are exceptions to the rule; in some animal species, the typical pattern of male–female differences is reversed, so that females are the pursuers and males the pursued. It is very revealing that in these species the usual pattern of parental investment is also reversed. For example, among pipefishes (close relatives of sea horses) males are smaller, less brightly colored, and more sexually coy than females, which are brightly adorned, physically aggressive, and sexually pushy. Although male pipefishes are clearly male (they make sperm, not eggs), they are uncannily like female mammals in their reproductive behavior: they nourish their offspring internally through a placentalike structure and eventually "give birth," with abdominal contractions reminiscent of a female in labor. Among pipefishes, the *males* provide the bulk of parental investment and display what might be called "feminine" traits, whereas the *females* compete among themselves for the males and might be called "macho."

Traditional sexual roles are also reversed among several species of South American birds known as jacanas. Female jacanas act like harem masters: they maintain large territories within which are several males, each with his own nest. The dominant female mates with all her males, lays eggs in their nests, and then leaves them with child-rearing duties. Each dutiful jacana "husband" incubates his egg clutch and provisions his young after they hatch. Not surprisingly, female jacanas are remarkably malelike, being larger than their mates, brightly colored, and sexually aggressive.

Insects provide some of the most telling exceptions, thus deepening the link between parental investment and sex differences. In certain insect species, the male produces an enormous protein-rich structure known as a spermatophore, which is eaten by the female. Noted nineteenth-century entomologist Jean-Henri Fabre seemed positively shocked as he recounted the sexual behavior of the cricket *Decticus albifrons*:

> The male is underneath, lying flat on the sand and towered over by his powerful spouse, who, with her saber exposed, standing high on her hind legs, overwhelms him with her embrace. . . . Have not the roles been reversed? She who is usually provoked is now the provoker, employing her rude caresses. . . . She has not yielded to him, she has thrust herself upon him, disturbingly, imperiously. . . . Master Decticus is on the ground, tumbled over on his back . . . and soon, from the male's convulsive loins is seen to issue, in painful labor, something monstrous and unheard of, as though the creature were expelling its entrails in a lump.

This monstrous lump is the cricket's spermatophore, a gelatinous package containing both sperm and nutrients. Because a spermatophore's nutrients are vital to egg production, male crickets often invest more in their young than do females. And this leads to dramatic role reversals: males take on sexual and social behaviors that are typically female in most other species, whereas females—like Madame *Decticus*—are aggressive and sexually pushy.

When do exceptions prove the rule? When they turn out, on closer inspection, not to be exceptions after all. Pipefishes, jacanas, and spermatophore-producing insects bolster the case for parental investment being a major player in the league of sexual strategies.

The Battle of the Sexes?

Although sex brings together two individuals that have a common goal—reproduction—the two sexes are not always in perfect, loving communion. Males and females each have a distinct agenda, and both have evolved not only to make the best deal possible for themselves but also to counter any tactical moves by the other.

In some species, males are reproductively competitive with the very females they want to inseminate. William Rice, an evolutionary geneticist at the University of California at Santa Cruz, has found, for example, that male and female fruit flies are engaged in an intense biochemical arms race, deploying strategies more competitive than cooperative. A variety of chemical tricks have evolved in both sexes. For example, the male's semen is laced with a protein that encourages the female to increase her rate of egg production while dampening her sexual appetite for other males. Should she mate again, another protein in the first male's semen will kill rival sperm but poison the female in the process. The female, however, produces her own chemical cocktail containing substances that eliminate some of the sperm in her body or act as an antidote to her mate's toxic proteins.

Working in the laboratory, Rice created two fruit fly populations: males that would evolve with each generation and females that were prevented from evolving by the design of the experiment. In only forty-one generations—a mere blink of the evolutionary eye—the males evolved ways of getting females that had already mated to remate with them while also preventing such females from mating with other males. As a result, females—which were evolutionarily disarmed in this particular war of the sexes—experienced a marked decline in life span because, Rice suspects, they were unable to evolve counterstrategies to toxic changes in the males' seminal fluid. Thus, it seems that each individual is out to make the best of his or her biological circumstances, even to the detriment of the mate.

What Do Women Want?

What competing is to males, choosing is to females. Among females, the propensity to be nasty and brutish and to beat one another-over the head rarely occurs. Females can afford to be—in fact, *must* be—demure and choosy. Why? Because the average female has too much at stake to squander her limited reproductive resources on just any male. For females, success is generally achieved by reproducing *well*—that is, by making successful offspring rather than by having a large number of mates. The female emphasis is on quality, not quantity.

Within our own species, for example, eggs (like those of most mammals) are a limited commodity; each one is a VIP that makes a special (usually solo) guest appearance, with a new egg appearing about every twenty-eight days during a woman's reproductive life. Once fertilized, the egg must then be nurtured internally for nine long months, and the offspring must be nursed for many months thereafter (at least in traditional societies). It is safe to say that at almost every stage children demand more from their mothers than from their fathers. Thus, it makes strategic sense for women to leave the fireworks, the persuasion, and the violence to men, who are competing among themselves for access to them. For good evolutionary reasons, females really are the gentler sex.

We do not mean to imply that women are not competitive with one another; they are, but not so directly or blatantly as men. Women tend to compete by focusing on themselves, by striving to look more attractive, seductive, desirable, and youthful than the competition, not by intimidating them or defeating them in battle. A woman who coifs her hair just so, flattens her tummy at the gym, and shops for a sexy dress may not be thinking, "I'm doing this to be more attractive than so-and-so." More likely, she is thinking, "I want to look really good so he'll notice me." The more resources a man has, the more likely women are to compete for his attention, a fact that explains why male rock stars, for example, are sometimes mobbed by their female fans.

Studies show that when it comes to exercising choice, females are drawn most to males that possess one or more of the following evolutionary "goods": good genes, good behavior, and good resources. Among humans, the captain of a football team or a man in uniform who is physically strong and capable might be viewed as having good genes; a churchgoing, devoted big-brother type might be a paragon of good behavior; and a millionaire businessman might be the embodiment of good resources. Simply put, the more of these qualities a male possesses relative to his competitors, the more attractive he is to females.

Good Genes. Because good genes are impossible to detect directly, females evaluate the genetic strength of their suitors on a number of grounds. Elephant seal cows, for example, vocalize loudly if mounted by a subordinate male, an act that brings the dominant male waddling over to displace the subordinate bull. In this way, the female ensures that her offspring will carry "good" genes.

Females of many species follow a similar strategy in looking for Mr. Goodgenes, as when a falcon engages in a difficult aerobatic pas de deux with her suitor, thereby testing his flying abilities and his likelihood of fathering offspring that will also be competent in aerial pursuits. A human parallel comes immediately to mind:

> He floats through the air with the greatest of ease,
> the daring young man on the flying trapeze.
> His movements are graceful, all girls he does please,
> and my love he has taken away.

Swedish biologist Malte Andersson studied female choice in the African long-tailed widowbird. The brightly colored males of this species, as the name implies, have spectacular—if silly-looking—tails measuring nearly two feet long. (In contrast, females of this robin-sized bird are dull colored and have short tails.) In a simple but elegant experiment, Andersson snipped the tails of some males and then glued the extra feathers to the tails of others. The result: females showed a distinct preference for the males with extended tails and snubbed those whose tails were unnaturally short.

Another Scandinavian researcher, Anders Møller, studied barn swallows, birds that—unlike the polygynous long-tailed widowbirds—are monogamous. Barn swallows' tails are not especially long, but they are notably forked. Møller found that females preferred males whose tails were deeply forked, but he also discovered that mated females would "cheat" on their mates by copulating with a deeply forked male if one came along. It seems female barn swallows that cannot get the male of their dreams as a full-time mate will nonetheless try to get his genes for their offspring. In so doing, they increase their chances of producing sons that have deeply forked tails and thus are more likely to enjoy reproductive success as adults. Biologists refer to this as the "sexy son" preference—that is, a female fondness for males whose offspring are likely to be sexually attractive to the next generation of females. (A human analogy would be a married woman who sought sexual liaisons with, say, an Olympic athlete or a Nobel laureate, unbeknownst to her husband, in the hope of producing a child who inherits her lover's desirable traits.)

Some biologists have recently proposed that the elaborate sexual ornamentation of male birds—bright feathers, wattles, combs, and so on—evolved not in direct response to female choice but as cues that the bearer is in good health. A survey of 109 species of songbird, for example, found that bright colors and fancy feather and song patterns are accurate reflections of disease resistance. Moreover, these traits are most elaborate in environments in which there are relatively large numbers of dangerous parasites. According to this view, some of these traits evolved as indicators of disease resistance, and thus of genetic quality. Either way, the basic picture remains unchanged: females are choosing males who possess good genes.

Good Behavior. The second evolutionary good much desired by some females is good behavior, specifically the kind that translates into parental care. Female Hawaiian damselfish, for example, prefer males that appear to be good egg guarders. They evaluate the males' prowess as would-be egg defenders during courtship, when males swim vigorously toward anything—whether female damselfish or potential predator—that approaches. Studies show that the males exhibiting the most impressive displays are the ones most likely to mate.

Not all promises of good behavior, however, are kept. Consider the case of the pied flycatcher. Among these small European birds, monogamous pairs rear offspring, with both males and females bringing insects to their ravenous young. But the male flycatcher frequently mates with a second female, which must rear her offspring alone because the male remains parentally committed to his first "wife." In such cases, the bigamous male locates his secondary territory some distance from the primary one, apparently to reduce the chances that the second female will discover female number one. Although some females clearly are duped in this way, the smarter ones do not mate without relatively long "engagement periods," which may give them time to discover whether their betrothed is hiding another female in the bushes somewhere. (The possibility also exists that secondary females are simply making the best of a bad situation if unmated males are in short supply or the remaining bachelors have poor territories.)

Incidentally, it is interesting to note that lengthy courtship and engagement periods are characteristic of human beings, a species in which both sexes are expected to make a substantial commitment to the rearing of offspring. The more time that passes between meeting and mating, the greater is the opportunity to check out the intentions, capabilities, and prior commitments of a prospective mate.

Good Resources. Of the three evolutionary goods females seek, a future mate's resources are the most apparent and, to many, the most important. From insects to primates, females prefer wealthy males. Indeed, many females drive a hard bargain, offering themselves only to males that proffer concrete proof of their endowment, say, in the form of food offerings, high-quality territories, and so forth. Among humans (as we describe at length in chapter 3), most women consider a great catch to be a wealthy man, someone such as a surgeon or a top executive, with the resources that allow for a luxurious lifestyle.

A male's resources are seen as so desirable, in fact, that in some animal species, females appear to practice prostitution, offering sex in return for goods. Among purple-throated Carib hummingbirds, males defend territories containing flowering trees, aggressively driving off other hummingbirds that attempt to feed there, with one exception: females that copulate with the territorial males are allowed to stay. Zoologist Elizabeth Gray, while a graduate student at the University of Washington, noted similar behavior among female red-winged blackbirds. If such a female is relatively poor—that is, if her territory doesn't offer much in the way of food—she may "cheat" on her mate by copulating with a neighboring wealthy male, which in turn permits her to forage undisturbed on his territory. Among bonobos (pygmy chimpanzees), females commonly approach males that are eating something desirable—a bit of meat or a choice piece of fruit—and wiggle their behinds in a sexual solicitation. While the male mounts, the female grabs the food. The details may differ, but there can be no doubt that similar bargaining occurs in our own species as well.

Of the three chosen goods, the first two—good genes and good behavior—are the most easily faked and, thus, the least reliable. There will always be males who try to present themselves as more than they are and put on a short-lived sexual display that is more show than substance. Good behavior can also be faked, as in the case of the bigamous pied flycatcher or the Hawaiian damselfish, the latter of which defends his territory with greater vigor and reliability when females are watching than when they are not.

Faking resources, however, is considerably more difficult. A courting male may pretend to be a paragon of genetic quality, monogamous fidelity, or devoted paternal attentiveness, but he will have trouble misrepresenting the quality of his real estate, the reliability of his nesting site, or the calories in a proffered insect, fish, or mouse. An old advertisement once intoned, "Promise her anything, but give her Arpege." Words are cheap, but it takes money—and the willingness to spend it—to buy expensive perfume.

Resource-rich males of nearly every species are remarkably attractive to females, at a level that often transcends conscious awareness. As part of courtship among humans, as discussed in the next chapter, men commonly ply their dates with flowers and seek to impress them with fancy cars, expensive clothes, and elegant restaurants often expecting sex in return. Rare is the man who doesn't have sex as a desired (if unfulfilled) goal at the end of a date. Women, in turn, send countermessages. By acquiescing, a woman may be signaling that the man has what it takes to win her affection; by refusing, she may be saying he does not or simply that she isn't sure, or she may be saying that her own prospective investment is so valuable that it cannot be obtained quickly or cheaply. Such women are exercising their choice, waiting for the right man—with the right combination of good genes, good behavior, and good resources—to come along. When it comes to female choice, it should not be surprising that humans are at least as adroit as damselfish or flycatchers. The details differ, but the patterns cry out to be acknowledged.

These accounts only scratch the surface of a rich vein of biological theory and observations about male–female differences among all living things as well as similarities between animals and human beings. In the chapters that follow, we shall return regularly to the world of evolutionary biology to tell of animals that illuminate the human condition. Such analogies do not, of course, imply that because bower birds, for example, use bright objects to attract mates or bellbirds call loudly during courtship, human females have some instinctive fondness for costume jewelry or necessarily swoon over males with deep, melodious voices. Rather, the animal kingdom provides examples of *general principles* that apply to human beings as much as to other creatures.

The results are undeniable: for all its elaborate emotional, symbolic, and cultural filigrees, sex—even among humans—is fundamentally driven by biology.

CHAPTER 3

Sex

THERE ARE ONLY the pursued, the pursuing, the busy and the tired.

— F. Scott Fitzgerald, *The Great Gatsby*

*H*uman beings are unusual, perhaps unique, in the animal world, in having liberated sexual behavior from reproduction. Few other animals bother with nonreproductive sex, and none have endowed it with as much complexity as have human beings. Not only do men and women cherish sex for its own sake, but some make it a religious ritual, an art form, an endurance sport, even a profession. Among humans, sexual intercourse means more than just making babies; it is an instrument of human communication, expressing the positive emotions of love, intimacy, excitement, and pleasure as well as some negative ones, including dominance, aggression, anger, hatred, neediness, and humiliation.

People occasionally think about reproduction while having sex, but most of the time they are otherwise engaged, worried more about being interrupted by children than about making children. But baby-making is nonetheless there, lurking beneath the lust, polishing and giving ultimate direction to erotic inclinations. Still, as any infertile couple can attest, few things are as unerotic as intercourse dutifully performed in the hope of fertilization.

Sexual relations tell us a lot about the evolution of fundamental differences between males and females, men and women. As described in chapter 2, natural selection tends to favor males that sleep around more or less indiscriminately, as long as some of those liaisons give rise to offspring and as long as the cost (in terms of physical output, time, risk, and so on) is not prohibitive. Virtually everywhere, men want sex more than women do and are willing to coerce, beg, bargain, and pay to get it. Rare is the woman who expresses a desire for indiscriminate sex. This is not to suggest that women lack a sex drive; rather, women are turned on by a different, and more conservative, pattern of sexual opportunities and relationships.

In this chapter, we investigate male–female differences, focusing almost exclusively on men and women. As we shall see, the gender gap in this arena reflects evolution's handiwork, with men turned on by sexual variety, responsive to quick forms of stimulation, and apt to be pushy. We also look closely at what turns women on, as well as three aspects of female sexuality—orgasm, concealed ovulation, and menopause—that, for all their importance, remain shrouded in mystery. Finally, we examine homosexuality to find what it reveals about the biology that underpins our sexual lives.

What Turns Men On

"Among all peoples, everywhere in the world," concluded noted sex researcher Alfred Kinsey and his colleagues, "it is understood that the male is more likely than the female to desire sexual relations with a variety of partners." Why? Because promiscuity has different biological consequences for males and females. Recall that a woman's parental investment is concentrated in a comparatively small number of offspring, each requiring considerable attention during the first few years of life. It is therefore to a woman's evolutionary advantage to choose the best partner she can (one who offers the right combination of evolutionary goods) rather than mate with just anyone. In contrast, men who "play the field" tend to enhance their reproductive fitness—except for those unfortunate few who are caught in the act and killed by an enraged husband. Overall, the net balance shows that when it comes to sexual variety, the two sexes differ markedly in evolutionary payoff.

For men, the most common sexual goal is to "score," just as in baseball, football, or basketball. (It is probably no coincidence that locker-room conversation often revolves around sexual conquests.) For women, the goal is both more subtle and more complex; a woman is less likely to keep track of the number of her sexual liaisons than is a man, less inclined to record sexual victories. Men go for quantity, women for quality.

Variety

A story is told in New Zealand about the early-nineteenth-century visit of an Episcopal bishop to an isolated Maori village. As everyone was about to retire after an evening of high-spirited feasting and dancing, the village headman—wanting to show sincere hospitality to his honored guest—called out: "A woman for the bishop." Seeing the scowl of disapproval on the prelate's face, he roared even louder: "*Two women* for the bishop!"

The Maori headman must be forgiven for his rather acute appreciation of male sexuality. In itself, the "gift" of a woman for the night was an acknowledgment of the strong cross-cultural male desire for sex. The revised offer of two women reflected the widespread male desire to seek sexual variety. This propensity among males for variety, common among other mammals and birds as well, is sometimes referred to as the "Coolidge effect."

As the story goes, President Calvin Coolidge and his wife were separately touring a model farm during the 1920s. Mrs. Coolidge noticed a large group of chickens associated with a lone rooster and commented, "He must be kept quite busy." After further discussion, she suggested that the rooster be brought to the president's attention. Accordingly, when the presidential party arrived at the new house, the guide announced, "Mrs. Coolidge wished me to point out that our single rooster copulates many times each day."

"Always with the same hen?" asked the president.

"No, sir."

"Well," Mr. Coolidge acidly replied, "tell *that* to Mrs. Coolidge."

The phenomenon of male promiscuity has been widely studied, especially among domesticated species, and has been found wherever sought: among sheep, cattle, cats, rats, and pigs. A ram, for example, left to copulate freely with the same ewe, will do so several times and then stop, apparently satiated. If he is then presented with a different ewe, his sexual enthusiasm returns until once again, his interest wanes. When copulating repeatedly with the same female, a ram will initially ejaculate after about two minutes. Intervals between ejaculation then widen to nearly three minutes, then five minutes, then fourteen minutes, and by the fifth exposure, nearly seventeen minutes. By contrast, if this same ram is provided with five opportunities to copulate but each time with a different female, his excitement will be such that he always ejaculates in two minutes or less. A new ewe makes a new him.

We know that male arousal of this sort is not triggered by simple removal and return because if the original female is taken away and then reintroduced (instead of being replaced with a new one), the male's sexual response is not comparably rejuvenated. The key is exposure to a *new* female. As another test, the original male can be replaced with a new one: the replacement male copulates enthusiastically with the original female, which is "new" as far as he is concerned. Thus, the waning sexual activity of the first male does not occur because the female, having copulated previously, is somehow less appealing to males generally or is no longer interested in sex. Sexual variety is what stimulates the males.

This phenomenon was known long before the modern science of animal behavior. "I have put to stud an old horse who could not be controlled at the scent of mares," wrote sixteenth-century French essayist Montaigne. "Facility presently sated him toward his own mares: but toward strange ones, and the first one that passes by his pasture, he returns to his importunate neighings and his furious heats, as before."

Human beings, too, are susceptible to the Coolidge effect. Men—even men who are deeply in love with their wives—are almost universally excited by the prospect of having sex with someone new. Few men would find it easy to refuse a naked, willing, and attractive woman who was suddenly and magically whisked into their bed. Yet after indulging in the illicit encounter, they might then repent, wonder what came over them, and profess undying love to their mate . . . and mean everything they say.

There is a large literature commenting on the Coolidge effect and the tendency for men to equate monogamy with monotony. Even Lord Byron wondered, "How the devil is it that fresh features / Have such a charm for us poor human creatures?" Similar sentiments were described by an African Kgatla tribesman in referring to sexual intercourse with his two wives: "I find them both equally desirable, but when I have slept with one for three days, by the fourth day she has wearied me, and when I go to the other I find that I have greater passion; she seems more attractive than the first. But it is not really so, for when I return to the latter again there is the same renewed passion."

When a man experiences diminished sexual desire for a familiar partner, it is not because he has said to himself, "I am limiting my reproductive potential by copulating exclusively with this woman, whom I may already have fertilized or am likely to fertilize at some future time." All he knows is that he is less excited at the prospect of sex with a familiar lover than with a new partner.

Similarly, a man indulging his penchant for sexual variety, or fantasizing about it, is unlikely to be thinking, "Ah! An opportunity to increase my reproductive fitness!" In fact, he may use a condom in a conscious effort to avoid reproducing. Nor is he likely to exclaim, "Hooray! A chance to indulge my male fondness for sexual variety!" More probably, he will either think "This is great," or—cognitive creatures that we are—place ethical considerations above his biological urges and choose to say "no." In either case, he is following a path established long ago by the forces of natural selection, with reproductive fitness being the ultimate source of his sexual energy.

Beyond acknowledging the fact that sex with a new partner produces a certain excitement, biologists can only guess at the underlying proximate mechanism. Brain cells and neurochemicals are known to become desensitized following repeated stimulation. Some have suggested, therefore, that after prolonged sexual association (perhaps weeks, months, even years), brain cells—male brain cells in particular—simply become saturated with neurotransmitters or resistant to them. According to one admittedly speculative account, "If you want a situation where you and your long-term partner still get very excited about each other, you will have to work at it because in some ways you are bucking a biological tide."

A Sense of Newness

On some level, all committed couples recognize the need to maintain a sense of newness in their relationship. Couples who maintain fulfilling sex lives over many years claim to have a lot of creativity when it comes to intercourse. They describe experimenting with positions, finding new settings, changing the timing, props, and so forth. Those who have a decidedly unimaginative approach (every Saturday night, with the man *always* on top, for example) tend to be the most bored with their love lives. Some monogamous couples seek help, looking to sex therapists and others for various ways to help restore passion to their relationship.

The more sexually oriented women's magazines are full of how-to articles replete with tips for avoiding sexual burnout. A random sample of cover lines from recent issues offers some typical fare: "Giving Him a Sexual Night to Remember"; "When Couples Stop Making Love"; "How to Turn the Heat Back On"; "Your Sex Life: Five Ways to Rock Your World"; and "Sexual Style: When Is It Time for a Change?" Obviously, there are a lot of couples striving for better sex; if this topic didn't sell magazines, it wouldn't be prominently displayed on the covers.

With so many couples turning to lingerie to infuse a sense of newness into their relationship, some might argue that the lingerie industry exists solely to cater to men's desire for variety. To a man's eye, a sexy nightie can transform a woman into an erotic, desirable lover. One moment she is the embodiment of everyday life; the next moment she is a woman transcended, her husband's fantasies come true. In fact, when a man gives his lover or wife a sexy teddy for her birthday (or anniversary or Valentine's Day), how often does she chide, "Is this for me, or is it really for *you*?" Although many, perhaps most, women take pleasure in wearing something slinky and erotic, lurking behind their enjoyment is the knowledge that seductive clothing makes them more attractive to their lovers. At some level, a woman also knows that a satisfied lover is more likely to be a faithful one.

Having said that, we have to admit that the biological tide is a strong one. Like it or not, most men thrill at the prospect of a new lover. For example, a survey was taken of paired adults—those in monogamous relationships—who complained of not getting enough sex. Of those, 62 percent of the men (only 37 percent of the women) said they would prefer "sex with someone other than their spouse or steady partner."

Many people claim that differences in sexual assertiveness reflect societal values, with men encouraged to be philanderers and women urged to be chaste. But such differences occur in every human society, are found among other species, and occur despite pressures exerted on men to be more sexually responsible and on women to be more sexually aggressive, at least within the context of a monogamous relationship.

We do not mean to suggest that men don't love, or even that they aren't capable of monogamy. However, most men simply do not equate sex with love to the extent that most women do. Anthropologist Donald Symons suggests that in a marriage, "women *give* sex for love, while men *give up* sex for love." As the famous team of sex researchers led by Alfred Kinsey pointed out:

> Most males can immediately understand why most males want extramarital coitus. Although many of them refrain from engaging in such activity because they consider it morally unacceptable or socially undesirable, even such abstinent individuals can usually understand that sexual variety, new situations, and new partners might provide satisfactions which are no longer found in coitus which has been confined for some period of years to a single sexual partner.

Nevertheless, it is a fascinating irony that although men stand to gain more—in terms of offspring—from multiple copulations, women are physiologically capable of having more sex than men. In his *Letters from the Earth*, Mark Twain had great fun with this paradox:

> Now there you have a sample of man's "reasoning powers," as he calls them . . . in all his life he never sees the day that he can satisfy one woman; also, that no woman ever sees the day that she can't overwork, and defeat, and put out of commission any ten masculine plants that can be put to bed to her. He puts those strikingly suggestive and luminous facts together, and from them draws this astonishing conclusion: The Creator intended the woman to be restricted to one man.

Now if you or any other really intelligent person were arranging
the fairnesses, and justices between man and woman, you would
give the man a one-fiftieth interest in one woman, and the woman
a harem. Now wouldn't you? Necessarily, I give you my word, this
creature . . . has arranged it exactly the other way.

Twain's Devil, the narrator of this excerpt, is absolutely right: one
man is less capable of sexually satisfying many women than one woman
is of satisfying many men. Nonetheless, from an evolutionary perspective, it is more logical for one man to mate with multiple women than
for one woman to mate with several men.

More Than One

Accordingly, polygamy (more appropriately called polygyny) has been
permitted by law and custom by the great majority of the world's people and, as far as we can tell, throughout nearly all of human history.
Because females roughly equal males in number, not all men can have
more than one wife. But it is clear that, historically, polygyny was a desired goal of most men in most societies, just as it is today wherever it
is legal.

In modern-day Taiwan, bigamy is common and culturally accepted,
especially among the wealthy. A movement is currently under way in
Russia—particularly supported, as one might expect, by many newly
rich capitalists—to institutionalize polygyny. In the United States, religious groups such as the Mormons once practiced polygyny, although
the custom was limited to elite males, that is, the relative few who could
afford to support multiple wives. "When it comes to polygyny," concludes Duke University anthropologist Weston La Barre,

> the cases are extraordinarily numerous. Indeed, polygyny is per-
> mitted (though in every case it may not be achieved) among all the
> Indian tribes of North and South America, with the exception of
> a few like the Pueblo. Polygyny is common, too, in both Arab and
> Negro groups in Africa and is by no means unusual in Asia and
> among Pacific islanders. Sometimes, of course, it is culturally lim-
> ited polygyny: Moslems may have only four wives under Koranic
> law—while the King of Ashanti in West Africa was strictly limited
> to 3,333 wives and had to be content with this number.

A look at other species reveals that monogamy is, in fact, rare. Among nonhuman primates, it occurs in only gibbons, siamangs, and marmosets. Among other mammals, monogamy has been described in some species of the wild dog family and among beavers, muskrats, dwarf mongooses, Asiatic clawless otters, elephant shrews, a few species of bat and seal, the reedbuck, and two small species of antelope. That's about it. This may seem like an extensive list, but it represents only a tiny fraction of the approximately 4,000 species of mammal.

Having multiple wives presents a man with one overwhelming evolutionary advantage: children. Assuming he has sexual intercourse equally often with each of his wives and that all are equally fertile, a polygamist with four wives might easily produce four times as many children (all with his genes) as a monogamist.

Is there any advantage to being a harem master's wife? On the one hand, a polygynous wife must share her husband's resources and parenting assistance with his other wives. But on the other hand, successful harem keepers tend to be relatively rich and powerful, so harem wives have high standards of living. "As my friend Prince Akiki Nyabonga of Uganda puts it," notes anthropologist La Barre, "a woman can hold her head up more as one of the wives of a man of substance than she could if she were the only wife of a poor, second-rate, monogamous husband."

This somewhat patriarchal view may well be true, although studies of people as widely separated as the Shipibo of the Peruvian Amazon and the Mukogodo of Kenya suggest that *women* in polygynous marriages often have fewer children, on average, than their monogamously mated counterparts. Still, one can assume that their children inherit not only their father's genes but also some of his resources and thus tend to be better off than children of monogamous parents.

Such a view is substantiated by animal research, notably a long-term study of red-winged blackbirds. Gordon Orians of the University of Washington found that females of this species often choose to be the second, third, or even fourth mate of a polygynous male rather than the monogamous helpmate of an erstwhile bachelor. Not surprisingly, these successful polygynous male blackbirds are "wealthy," occupying highly desirable territories, so females gain by being associated with a harem.

But for most men, multiple wives are more the stuff of dreams than reality. Even when polygyny is encouraged, only a small handful of men actually have the resources to support more than one mate. Joseph Smith, founder of Mormonism, who is said to have had as many as forty-six wives over the course of his life, was certainly the rare exception.

Typically, men in polygynous societies accumulate wives as they get older, in the process gathering power, wealth, and status. In fact, one reason—perhaps the main one—why men in polygynous marriages strive to accumulate power, wealth, and status may be that doing so helps them accumulate wives. In some cases, this connection may be conscious; but in any event, reproductive success is probably the ultimate evolutionary rationale behind such strivings.

Another common pattern is for men to have serial wives, as in the past, when young women more frequently died in childbirth. Monogamous widowers typically remarried and then fathered more children, creating stepfamilies similar in size and complexity to those in Western society that are currently created by divorce. Overall, men who lose their wives—whether by death or divorce—tend to remarry at much greater rates than do women who lose their husbands.

Visual Stimuli

As almost any couple can attest, men are more readily aroused than women. A sexy word, a provocative image, and a man's sexual motor is revved, whereas women typically take longer to reach the same level of excitement.

Picture the following scene: a man and woman have just spent the evening together and are now in the privacy of her apartment. She turns to him and says, "This has been a lovely evening: wonderful conversation, dinner, and movie. I've really enjoyed your company and getting to know you, and, well, I wonder if maybe you'd like to get to know me better, too." With this, she slowly unbuttons her blouse and then starts to unhook her bra. All the while, the man watches: fascinated, excited, delighted. Then she uncovers one breast and he . . .

What is he likely to do? Slap her across the face? Beg her to stop and put her clothes back on? Scream for help? Dial 911? Most of us can write a plausible ending to this steamy—if unlikely—encounter, and whatever the precise details, it will probably involve increased sexual intimacy between the two people in question. The sort of female behavior described here would be a definite turn-on for most men.

Now, run through the scene again, this time reversing the roles: after making a similar statement, the man unzips his fly, revealing his erect penis. Most likely, the woman reacts with something less than delight, fascination, and sexual enthusiasm. Rather than excited she may well be repulsed, almost certainly by the man's social inappropriateness and, perhaps, by the penis itself. In any event, she is not likely to be aroused by the visual image of his genitals.

According to Kinsey and his collaborators, "Many females consider . . . male genitalia ugly and repulsive in appearance, and the observation of male genitalia may actually inhibit their erotic responses." In fact, blatant male sexual display—as in the case of "flashers"—is widely seen as a pathetic aberration, certainly not as sexually attractive. In many traditional cultures, male genital display conveys threat rather than enticement.

By contrast, nearly every heterosexual man is excited by exposure to the intimate anatomy of an attractive woman. If a woman displays her genitals, no matter what the culture, the act is seen as an invitation, never a threat. In a recent issue of *Cosmopolitan*, a man describes a rewarding sexual experience initiated by a woman: "We were on our way to dinner, right outside her apartment door, and she lifted her skirt. I bent over her and we had sex—right in the hallway! The element of surprise and spontaneity made it incredibly exciting." We suspect, however, that the encounter was more fantasy than reality. (And male fantasy, at that.)

Many men also assume, often incorrectly, that women will be as stimulated by erotica as they are. "Most males," wrote Kinsey and his colleagues,

> find it difficult to comprehend why females are not aroused by . . . graphic representations of sexual action, and not infrequently males essay to show such materials to their wives or other female partners, thinking thereby to arouse them prior to their sexual contacts. The wives, on the other hand, are hurt to find that their husbands desire any stimulation in addition to what they, the wives, can provide, and not a few of the wives think of it as a kind of infidelity which offends them.

Lorraine was, to put it mildly, unhappy on discovering her husband, Joe's, secret collection of erotic magazines, and Judith spent several months helping the two put their marriage back together. Lorraine was in turmoil; to her mind, she and Joe were good, solid, churchgoing citizens, as far from being "perverts" as could be imagined. The couple were in their forties and had three teenage children who were doing well. Both Lorraine and Joe were Methodists who had grown up in the same midwestern state and both considered themselves moderate Republicans. In recent years they had sexual intercourse about once a week, on Saturday nights; neither had ever sought counseling or complained of marital problems.

When Lorraine accidentally discovered Joe's magazines in the tool shed, she fell apart. She was emotionally overwhelmed; her first response was to vomit and then burst into tears. By the time Joe came home from work, Lorraine was in shock. She tearfully claimed that he had lied to her, that she didn't really know him, and that perhaps they should get a divorce. Joe was dumbfounded: he readily acknowledged his "dirty little secret," but claimed it wasn't really all that unusual. Moreover, he said he planned to donate his collection to the library because by this point it had collector value and they could take a tax write-off! Although Joe was embarrassed, he couldn't make heads or tails of Lorraine's panic.

Lorraine, however, felt as if she had caught Joe having sex with another woman; she was inconsolable, with an acute sense of betrayal and outrage. Although Joe sheepishly acknowledged that he occasionally masturbated to photographs, he vigorously asserted that he had never had an affair, never intended to, and didn't know a man who had not masturbated to pictures at some time or another. Judith helped the couple understand their sex differences so that Lorraine could learn to accept Joe's apology, forgive him, and regain trust. Joe, meanwhile, needed to understand the impact of what seemed to him a minor deception. Happily, the couple also began reporting greater sexual satisfaction and intimacy.

Wives and girlfriends, chagrined to discover a hidden copy of *Playboy* or *Penthouse*, or troubled by their lover's interest in X-rated movies, would do well to realize that their husbands and boyfriends are not necessarily deviant or even sexually dissatisfied. Like the great majority of men, they are simply susceptible to stimulation. Whether in the form of films, calendars, cartoons, stories, or computer images, pornography is consumed primarily by men.

Women should also realize that few male viewers ever concern themselves with the personalities of erotic female models: their likes, dislikes, jobs, even their names. It would be almost inconceivable—if anything, a sexual turn-off—for most men to consider whether pornographic subjects have fathers, mothers, sisters, brothers, or children. Erotic images stimulate men to imagine what they would do to, or perhaps with, the posed women, not to imagine them in a family setting. To this extent, then, women need not be threatened by their husbands' fascination with titillating images. Perhaps they might even be grateful: erotica permits men to gratify their longings for "sex" with many different women, but without the dangers of venereal disease or troublesome entanglement.

As might be expected, women are rarely aroused by images of nude or partially dressed men. Several decades ago, Kinsey and his colleagues found that only 12 percent of the women they studied reported sexual arousal when shown photographs, paintings, or drawings of nude men and women. These results have been confirmed in a more recent study as well.

In other research, heterosexual college students were shown pornographic films depicting various combinations of lovers while measurements were taken of their sexual arousal (degree of genital engorgement, amount of sweating, elevation of heart rate). Women were consistently less aroused by the films than were men, although they were most turned on by heterosexual couples. Men, in contrast, were most turned on by group sex, and both men and women were attentive to the women in the films.

Just as interest in having sex with the same partner can diminish over time, interest in pornography and sexy pictures tends to decline after a certain threshold of exposure or when the allure of secrecy and novelty is removed. In the 1960s, scientists at the University of North Carolina (under the leadership of Morris Lipton, a leading psychiatric researcher and Judith's father) recruited a group of undergraduate men to determine what effect daily exposure to pornography would have on their sexuality. The students were required to view pornographic movies or materials for two hours every day over an entire month. During this month, the researchers studied the students' testosterone levels, their moods and feelings, and reports from their girlfriends about sexual behavior. Not a single student experienced an increase in sexual interest; in fact, the students reported increased sexual boredom, and their partners described reduced desire for sexual intercourse.

Although we agree that pornography is degrading and generally oppressive to women, we also believe strongly that the phenomenon is driven more by profit making than by a misogynist social agenda. As Bob Guccione, publisher of *Penthouse*, can attest, a lot of money is made by pandering to male susceptibility to female sexual images. We have great confidence in capitalism, at least to this extent: if women were as eager to pay for images of penises and scrotums as men are to pay for the sight of women's genitals, "boyie" pornography would assuredly be as prevalent as the "girlie" variety. What men get from pornography, women get largely from the romance novel because to a great extent, what men get from sex, women get from romance.

Any Woman Will Do

According to a survey, men are more than twice as likely as women to fantasize about group sex, more than twice as likely to fantasize about voyeuristic or fetishistic scenes, and one-third less likely to include their current steady partner in such fantasies. Women are nearly four times as likely as men to imagine having sex in romantic or exotic settings (islands, waterfalls, moonlit nights, and the like) and more than twice as likely to report never having had sexual fantasies at all.

In addition, men often fantasize about female sexual anatomy such as nipples and labia with almost clinical vividness, but women rarely report fantasizing about anatomical details such as penis size or shape or hairiness of a man's chest. Also, whereas men's fantasies are split evenly between active and passive sex (doing as well as being done to), women's are twice as likely to be about passive sex. This difference in men's and women's fantasies has been found cross-culturally, as true in Japan as in Great Britain.

Because women prefer romance to anatomy, it is highly unlikely that a book modeled after Madonna's *Sex* by, say, Brad Pitt, Keanu Reeves, or Robert Redford, would ever be a runaway success, yet a comparable book by another female sex symbol—Sharon Stone, Julia Roberts, or

Whitney Houston—is not at all difficult to imagine. It is noteworthy, however, that women's magazines with a distinct sexual content (*Cosmopolitan, Glamour, Elle, Self*) feature attractive young *women*—not unlike their male-oriented counterpart magazines, although their models wear considerably more clothing. Men look at a woman's photograph clad or unclad, and often imagine having sex with her; women, we suspect, like to check out the competition. Unlike men, who fantasize about having sex with the model, women typically fantasize about *being* her.

Consistent with this lower threshold for male sexual arousal, prostitution, like pornography, is a sexual service provided primarily by women (though sometimes by boys or men) for which men pay. Very rarely is it the other way around—women paying men—and for good reason: women who want sex can almost always get it for free. Natural selection has ensured that a ready supply of sperm donors will very likely be available, and generally willing to perform, if given the opportunity. It seems that women have only to indicate availability to evoke sexual interest on the part of some male—often more interest, and more quickly, than many women would prefer.

The more attractive a woman, the more likely she is to have many suitors. But virtually any woman is capable of finding a mate (with a distinction made between mate and spouse); she may have fewer choices than would a more attractive woman, but she will have choices nonetheless. An obstetrician friend tells us about a derogatory term, "mystery mom," that he sometimes hears in the labor and delivery unit of his hospital. The term refers particularly to heavyset, masculine-appearing women who pose a mystery because no one can imagine any man having had sex with them. Nevertheless, for every such woman, at least one man found her—if not desirable enough for marriage—attractive enough for a bout of sex. The take-home message here is that women, even unattractive ones, can always get sex, whereas similarly unattractive men are generally out of luck.

Men Who Are Superachievers

Some men, of course, are greatly desired by women, which might seem like role reversal except that such men nearly always possess, or appear to possess, exceptional resources. Once again, biology prevails: men who are especially wealthy, famous, or charismatic are pursued by women wherever they go.

A recent *Cosmopolitan* article claims "A profligate sexuality seems to be a perk of power in this society." We agree, but we think the article gets it only half right in suggesting that superachievers are supersexed. Although a superachiever may well have a higher level of testosterone than the average shoe salesman, what the article doesn't add is that powerful men often have multiple sexual liaisons because superachievers—married or not—constantly have women coming on to them.

Chief executive officers, rock stars, sports heroes, and political movers and shakers all describe being bombarded by seductive women. John F. Kennedy and Martin Luther King Jr. are just two charismatic leaders who are said to have had insatiable sex drives. More likely, they had inexhaustible sexual opportunities. Basketball star Wilt Chamberlain claims to have slept with approximately 20,000 women (give or take a few hundred, we presume). Even allowing for some exaggeration, sexual conquests at this level are a privilege enjoyed by only a few.

Hugh Grant, named sexiest movie star of 1995 by at least one magazine, shook celebrity gossip circles when he was arrested for having sex with a prostitute. What shocked everyone was not the notion of Grant consorting with a woman other than his girlfriend (a gorgeous cover girl, no less) but the fact that he had paid for something many women would gladly have given him. Why would he do such a thing? Our hunch is that he viewed the simple exchange of money for sex as less cumbersome than a tryst because virtually any woman who offered free sex might well have wanted more from him than money: future dates, affection, connections, even a child. More than a few movie stars have found themselves paying child support after brief liaisons that seemed, at the time, "just for fun." Hence, in trying to make a simple, low-parental-investment exchange of money for sex, Grant—for all his renown—was following a traditional male strategy.

Liking Bad Girls But Marrying Good Ones

Women who offer men sex in hopes of securing a more permanent commitment cannot afford to make themselves too available: sex is one way for a woman to demonstrate her love for a man, but if she is too "easy," she runs the risk of turning him off, at least as a potential long-term mate. In *Humboldt's Gift*, by Saul Bellow, the protagonist reveals a widespread male concern about a prospective mate who may be too good, that is, too experienced as a lover: "As a carnal artist she was disheartening as well as thrilling, because, thinking of her as wife-material, I had to ask myself where she had learned all this."

Men are supposed to like bad girls but marry good ones, as goes an old saying with a modern evolutionary ring. Despite the commercial success of the movie *Pretty Woman*, most viewers would agree that its plot, in which a millionaire falls for a prostitute, is highly unlikely. Indeed, the well-described tendency for men to divide women into two groups—whores and madonnas (the idealized Virgin Mary, not the pop star)—reflects men's contradictory desires: to have numerous sexual adventures and opportunities (with whores) but to marry someone relatively chaste (a madonna), who probably will not squander the husband's resources on another man's offspring. Such concerns are not unique to *Homo sapiens*: males of other species seem to prefer sexually chaste females as well. A study of ring doves, for example, revealed that males reject potential mates that demonstrate, by their sexual eagerness, that they have already been courted (and perhaps inseminated) by another male.

Weird Sex

Why is it that some gentlemen prefer blondes and others prefer enema bags, whips and chains, corpses, or young children? Almost without exception, weird sex is male sex. It is men, rarely women, who engage in sex with animals, who become obsessively dependent on underwear or other articles of clothing, who are turned on by violence or urine or masturbate with feces. Significantly, when women participate, they generally do so out of fear or for pay, with either the threats or the money coming from—you guessed it—men.

Such perversions are technically known as "paraphilias," the word deriving from the Greek *para* meaning "deviant," and *phila*, meaning "attraction." Paraphilias are defined as arousal in response to sexual objects or situations that are not part of normal arousal and that to varying degrees interfere with the capacity for affectionate, reciprocal sexual relations. Paraphilias generally involve an intense or even compulsive desire for sex with a nonhuman object or a nonconsenting partner, or for the suffering or humiliation of oneself or another.

To count as full-blown pathology, a paraphilia must be persistent and involuntary, as opposed to the occasional turn-on. Paraphilias include, but are not limited to, compulsive exhibitionism (exposure of one's genitals); fetishism (repeated use of nonliving objects, such as articles of clothing, to achieve sexual arousal); frotteurism (touching or rubbing of a nonconsenting person—for example, achieving ejaculation by rubbing against an unwitting fellow passenger in a crowded subway car); pedophilia (intense sexual urges directed toward children); masochism (achievement of sexual excitation by being humiliated, bound, beaten, or otherwise made to suffer); sadism (infliction of psychological or physical pain during sex); voyeurism (intense observation of others—often without their knowledge—as a means of sexual satisfaction); zoophilia (preferred or exclusive use of animals); telephone scatologia (obscene telephone calling), necrophilia (sex with corpses); partialism (exclusive focus on a part of the body, such as feet or ears); klismaphilia (sexual excitement associated with enemas); urophilia (sexual focus on urine); and coprophilia (sexual use of feces). Again, the striking fact about this unsavory array is that these acts are engaged in almost exclusively by men.

Sexual masochism, for example, occurs in about twenty males for every female; the other paraphilias are practically never diagnosed in females, although some cases have been reported. Even sexual harassment, which is predominantly practiced by men against women, can involve a degree of paraphilia. An example is seen in the recent case of a chief executive officer of a large corporation, who was found guilty of sexual harassment based on evidence that he kept the door to his executive washroom open while he used the toilet and required his secretary to retrieve notes he had hidden in his clothing, among other acts.

Why are men so susceptible to what we might label the P-cubed phenomenon: pornography, prostitution, and paraphilia? The most logical explanation is that natural selection has favored a relatively indiscriminate, shoot-from-the-hip sexuality among males. Granted that the primary reproductive strategy of men is not to copulate with every warm body but to invest parentally in the offspring of one or more primary mates, evolution has nonetheless favored those who are alert to the possibility of inseminating additional casual partners. With such itchy trigger fingers, men are likely to misfire or shoot wide of the mark, and with little cost to their reproductive fitness (at least in the days before DNA testing and mandated child support).

To some degree, paraphilias represent the desperation of relatively low-ranking individuals who have normal sex drives but lack the wherewithal to court successfully. For example, one of Judith's patients was a self-proclaimed sadistic paraphiliac. Caught painfully in a fear of both sexes, he developed an "addiction" to satanic images and black magic, engaging in blood rituals that he performed alone as an expression of rage at himself and the world. He masturbated to images of blood and bondage because he was afraid of anything more intimate and hated his shyness and everyone else for having more fun in life.

What Turns Women On

"Not tonight, dear," goes the familiar refrain. "I have a headache." Or "I'm too tired." Why do we automatically assume that the voice is female? Because usually, it is. But why? In most cases, it is women who hold back when it comes to sex, whereas men consistently want it. One woman speaks for millions when she says, "My husband always wants sex; he'd have to have one foot in the grave to say no." Not only do most women want sex less often than men, but they prefer to have it in a loving and romantic environment. When *Glamour* magazine asked its readers to describe the best sexual experience they had ever had, very few female respondents included anatomic details. Instead, the vast majority described the mood and setting and wrote of being pampered and made to feel special by their lovers. Women, it seems, want to be cuddled and coddled, not simply inseminated, whereas most men are perfectly content with the latter.

Even when a woman acquiesces, however enthusiastically, to a sexual encounter, she typically does so not for the sake of intercourse per se but to prove to the man that she loves him, to clinch their emotional closeness, or to reward him for prior attentions. In short, sex is most often something the man wants and the woman agrees to. "Among men," writes Donald Symons, "sex sometimes results in intimacy; among women, intimacy sometimes results in sex."

Although women can engage in sex while in an indifferent or even hostile frame of mind, merely enduring the act without any particular excitement or pleasure, most prefer a relaxed and safe environment, soft lighting and other touches to create the right mood, and conversation at the same time or, alternatively, perhaps complete silence or extra lubrication or genital stimulation with fingers, tongue, or vibrator in order to reach orgasm.

We speculate that women evaluate the suitability of potential mates at least partly on the basis of their sexual behavior. A thoughtful, considerate lover is likely to be a thoughtful, considerate husband. On the other hand, some men present themselves as more thoughtful and empathic than they really are, and some women are especially adept at detecting this deception. In Colette's novel *The Vagabond*, for example, a woman criticizes (and sees through) her admirer as follows:

> If he pretends, cunning as an animal, to have forgotten that he wants to possess me, neither does he show any eagerness to find out what I am like, to question me or read my character, and I notice that he pays more attention to the play of light on my hair than to what I am saying.

In contrast, Czech novelist Milan Kundera describes a canny male conquest:

> On that fateful day, a young man in jeans sat down at the counter. Tamina was all alone in the cafe at the time. The young man ordered a coke, and sipped the liquid slowly. He looked at Tamina. Tamina looked out into space. Suddenly he said, "Tamina."
>
> If that was meant to impress her, it failed. There was no trick to finding out her name. All the customers in the neighborhood knew it.
>
> 'I know you are sad,' he went on.
>
> That didn't have the desired effect either. She knew that there were all kinds of ways to make a conquest, and that one of the surest roads to a woman's genitals was through her sadness. All the same she looked at him with greater interest than before.
>
> They began talking. What attracted and held Tamina's attention was his questions. Not what he asked, but the fact that he asked anything at all.

Kundera understands, of course, that it is not sadness but skillful use of emotional language and caring dialogue that are the surest road to a woman's genitals. To a large degree, just as men are turned on by visual stimuli, women are especially influenced by talk. It is not that women don't look or men don't listen but rather that men look especially carefully and women listen especially carefully for cues that suggest a possible mate. Overall (recognizing that there are exceptions to this generalization), women choose their sexual partners carefully, and for good reason.

One at a Time

Women are consistently about as averse to sexual variety as men are drawn to it. As Kinsey pointed out, "Many females find it difficult to understand why any male who is happily married should want to have coitus with any female other than his wife." We submit that this is not simply because society has sought to repress female sexual desire (although it has, and for reasons that make biological sense) but because most women do not experience increased sexual desire when presented with a new, anonymous partner, with emphasis on the anonymous. A woman who awoke to find a naked man crawling into bed with her would be likely to bolt and call the police; a man who found a naked woman in his bed would be likely to stick around, thinking his ship has come in.

Again, women respond as they do almost certainly because at the ultimate, evolutionary level, a new partner *as such* is unlikely to enhance—and may well impair—a woman's reproductive success. Thus, women lack a comparable "Mrs. Coolidge effect." Certainly women are capable of engaging in sexual intercourse with new and different men—sometimes, as in the case of prostitutes, many different men in succession—but this is quite different from being inspired to do so by the very newness of the partner. Indeed, sexual variety itself is cited by prostitutes as one of the emotionally deadening aspects of their vocation.

On occasion, of course, women are unfaithful to their partners, though the rates of adultery among women are consistently reported to be lower than those among men. Almost never, though, do women seek sex with male prostitutes or men they don't know and are unlikely ever to see again. A study of adultery among American women further reinforces the view that their sexual motivation is different from men's:

"Extramarital sex . . . had far more to do with holding on to or obtaining a partner—with living in pairs, albeit sequentially—than with living in threes and fours, and at sixes and sevens."

Such findings do not surprise today's evolutionary biologists. Even psychoanalysts, with their different theoretical orientation, generally concur. Psychoanalyst Helene Deutsch had this to say about women: "Our impression is that the feminine woman in an overwhelming majority of cases is fundamentally monogamous. This monogamy does not necessarily require the exclusiveness of marriage or confine sexuality to one object for life. A woman may even change her love objects quite frequently; but during each relationship she . . . has a conservative need to continue [it] as long as possible."

Rewards of Intimacy

A study of couples on U.S. college campuses found that the duration between initial meeting and onset of sexual intimacy depends on various characteristics of the young women, including prior sexual experience, religious beliefs, and hopes for the relationship. In other words, the women were the ones in control, engaging in sexual relations only when they were ready. Interestingly, the characteristics of the young men were virtually irrelevant. Essentially all of them, whoever they were, wanted one thing: sexual intercourse.

Why is it that men are sex oriented and women are relationship oriented? Maybe sex is simply more rewarding for men; after all, intercourse nearly always results in male orgasm, whereas female orgasm is more "iffy." Yet even in a relative sexual paradise for women such as the Polynesian island of Mangaia, where multiple orgasms among women are the rule, men want sexual intercourse more than women do. On Mangaia, thirteen- and fourteen-year-old boys are instructed in sexual behavior by a designated, experienced woman. In addition to learning various positions for intercourse, they are taught techniques of foreplay including fondling, kissing and sucking the breasts, cunnilingus, and especially the desirability of inducing one's partner to have several orgasms before themselves ejaculating, that last act preferably timed to match their partner's climax.

Mangaian women expect orgasms from their partners and typically leave men who do not sexually satisfy them. Although intercourse on

Mangaia is not extraordinarily prolonged (lasting fifteen to thirty minutes), it far exceeds the Western average (variously estimated at from two to ten minutes) and presumably is very satisfying to women. Nonetheless, virtually always it is the Mangaian men who initiate sex; the women are more coy and reluctant.

Resources

To be sure, women, too, want sex and probably always have. But they are far more likely to link sex to romance or use sex as a means to an end (resources and relationship) than to pursue it strictly for carnal pleasure. For example, it is a sure sign that a man desires a woman sexually when he produces expensive gifts: long-stemmed roses, jewelry, clothing, theater tickets, or something else that is especially meaningful. Whatever the offering, men—especially those with resources—entice women with presents. And women often accept such gifts, either content with the simple exchange of sex for material goods or in hopes of receiving something more permanent from their resource-laden men. The fact that John F. Kennedy Jr. (at least before his marriage to Carolyn Bessette) was voted the "sexiest man in America" should come as no surprise. True, he is genuinely handsome and the only son of a charismatic former president, but it also matters very much that he is a multimillionaire and increasingly influential in American politics.

Suffice it to say, that men almost never give spontaneous or lavish gifts to women who are just friends. And although some couples do conduct a mutual, egalitarian exchange during courtship, in few cases do women seek the sexual attentions of men by showering them with gifts. Even among animals, sexual bribes are common and nearly always pass from males to females. A well-known case concerns insects known as scorpion flies, studied by University of Mexico zoologist Randy Thornhill. Female scorpion flies will not mate unless the males present them with a protein-rich nuptial gift—generally some smaller insect they have killed—before mating. Those females that hold out for wealthy, generous males lay more eggs, and male scorpion flies are thus rewarded for their gift giving.

Such one-sided "generosity" reflects a universal—although not usually acknowledged—truth: as with scorpion flies and most other species, the woman's gift is her body, that is, her parental investment.

(The man's proximate reward is his sexual gratification; his ultimate reward is the woman's potential for childbearing.) Feminist social philosopher Simone de Beauvoir agrees: "From primitive times to our own, intercourse has always been considered a 'service' for which the male thanks the woman by giving her presents or assuring her maintenance."

Pioneering anthropologist Bronislaw Malinowski reports that even among the Trobriand Islanders, often cited as an example of easygoing sexuality in which the line between pursued and pursuer is comparatively blurred, in the course of every love affair the man must constantly give small presents to the woman:

> This custom implies that sexual intercourse, even where there is mutual attachment, is a service rendered by the female to the male. . . . This rule is by no means logical or self-evident. Considering the great freedom of women and their equality with men in all matters, especially that of sex, considering also that the natives fully realize that women are as inclined to intercourse as men, one would expect the sexual relation to be regarded as an exchange of services itself reciprocal. But custom, arbitrary and inconsequent here as elsewhere, decrees that it is a service from women to men, and men have to pay.

We take Malinowsky's observations as accurate, although we differ with his interpretation. It is not simply "custom, arbitrary and inconsequent" that has decreed sex a service rendered by women to men but, rather, biology—consistent, farsighted, and consequential—that calls the tune.

Men also pay for sex in other ways, by proving themselves worthy through acts of bravery and heroism. According to one student of sexual anthropology:

> The underlying understanding is that women, even wives, grant sexual gratification to men in accordance with how well they fulfill their masculine roles. The traditional expression of this is contained in many heroic stories and songs in which the vigorous young man professes his love. The girl urges him to go out and demonstrate his courage and skill in fighting. He returns with trophies that prove his bravery, and she grants him his desire.

Modern-day heroes are similarly rewarded, as any sports star, life-guard, or decorated police officer can attest.

Making Choices

So why, if women can afford to be the choosier sex, do so many of them make such bad choices and end up married if not to cads, then to men who seem undeserving of them? The question is far too complex to be thoroughly addressed here, but we can provide some insights. To begin with, few "perfect" men (those with an optimal combination of genes, behavior, and resources) exist. As a consequence, most women—and men—must make compromises, settling for something less than the ideal mate. We contend that most women go for what seems best at the time: a man who appears emotionally sensitive or who has a strong sense of family values or a thick wallet. In fact, when choosing a husband, many women opt for a man who is particularly caring, even if not very attractive or smart, over someone who is more outwardly charming but also more self-absorbed.

When a friend of ours told her parents she was having second thoughts about her fiancé and was no longer sure she wanted to marry him, they responded, "But he's got such a good job!" To the parents, their prospective son-in-law's potential to be a good provider outweighed whatever flaws he had. The daughter, however, had other ideas and eventually left her fiancé, with no regrets.

Parental interests aside, most women would agree that a well-heeled man is attractive; the phrase "She made a good catch" almost always refers to a woman who marries someone with a good income. But most women would probably also agree that there is little pleasure in being married to a wealthy man who shows no interest in his wife, has extramarital affairs, and perhaps is unattractive to boot—unless, of course, he is *very* wealthy. However, there is also little pleasure in being married to a man who exhibits good behavior (someone, say, who is exceptionally faithful) or possesses good genes (an athletic physique or sharp intellect) if he sits around the house all day, doing little to provide for his family.

Occasionally a woman of higher socioeconomic status chooses a lower-ranking man who may be otherwise attractive though lacking in economic resources, but such relationships are rare and usually are fraught with problems.

One of Judith's patients, Jessica, typifies the frustration such women may feel. More educated than her former husband, Jason, who is a free-lance artist, Jessica earned three times his income. Until the couple had a child, their marriage was satisfactory, with Jessica feeling that her husband made up in personal warmth and artistic talent what he lacked in money. However, when she became pregnant and developed gestational diabetes, she found Jason's lack of financial resources troublesome and his artistic interests infuriating. "While I sit here with my swollen legs up on pillows," Jessica told him, "trying desperately to bring home the bacon, you paint your watercolors of birds and flowers, lost in another world!" The situation worsened after the baby was born and Jessica found herself supporting three people, with no time for herself and no respect for her husband. A divorce soon followed: "It's easier to take care of the baby myself than to watch that loafer exploiting me," she told Judith. "Tell him to get a job that pays."

Overall, a man who possesses marginal physical or intellectual charm is unlikely to be successful with women if he also lacks power and money. George Orwell wrote perceptively about this in a discussion of the plight of the English hobo, or "tramp":

> The result, for a tramp, is that he is condemned to perpetual celibacy. For of course it goes without saying that if a tramp finds no women at his own level, those above—even a very little above—are as far out of his reach as the moon . . . there is no doubt that women never, or hardly ever, condescend to men who are much poorer than themselves. A tramp, therefore, is a celibate from the moment when he takes to the road. He is absolutely without hope of getting a wife, a mistress, or any kind of woman except—very rarely, when he can raise a few shillings—a prostitute.

Female Sexuality

Perhaps because men fear ending up as sexual tramps—celibate and unloved—female sexuality has assumed immense importance in the minds of most males. It has also been vastly distorted by both sexes. At least one writer (in this case, an ardent feminist) argues that historically, female sexuality was naturally voracious, uninhibited, and unlimited. But then, sadly, "Primitive woman's sexual drive was too strong, too susceptible to the fluctuating extremes of an impelling, aggressive eroticism to withstand the disciplined requirements of a settled family life." Fortunately, there is no evidence to support such a claim. As noted psychologist Frank Beach points out, "Any male who entertains this illusion [that women are sexually insatiable] must be a very old man with a short memory or a very young one due for a bitter disappointment." Or, as Donald Symons put it, "The sexually insatiable woman is to be found primarily, if not exclusively, in the ideology of feminism, the hopes of boys, and the fears of men."

Men, in their befuddlement, have had a hard time seeing female sexuality for what it is, consistently either over- or underestimating it. Thus, women have often been portrayed as either rapacious and insatiable or lacking sexual desire altogether. At one time, Talmudic scholars entertained such an overblown estimate of women's sexuality (and society's responsibility to repress it) that widows were forbidden to keep male dogs as pets! At the other extreme, an influential nineteenth-century Victorian physician, Dr. William Acton, announced that "the majority of women (happily for society) are not very much troubled with sexual feelings of any kind. What men are habitually, women are only exceptionally."

Even today, a woman's sexuality is often measured by her ability to arouse desire in a man, rather than her capacity to experience it herself. These distortions—by men—arise largely because they view female sexuality as threatening. Not only does a fully sexual woman evoke deep fears of inadequacy in some men (especially the fear of being unable to satisfy such a woman), but a woman's sense of sexual freedom also threatens a man's confidence that his wife's children are his own. In short, men are often comforted by the myth of the unsexed woman even if deep in their hearts, they know it is a lie.

Mysteries of Womanhood

Freud once wrote that the psychology of women was a "dark continent," impossible to understand. Evolutionary biology, we believe, shines a bright light on the psychology of both sexes. Nonetheless, some aspects of female sexuality remain mysterious, although there is reason for confidence that they will soon be illuminated.

The Female Orgasm

Much ink has been spilled over the question of whether human females are unique in experiencing orgasm. In the past, it appeared that they were; more recent evidence, however, suggests that other female primates experience orgasm as well, although to various degrees. For our purposes, the uniqueness of the female orgasm is less important than the fact that unlike ejaculation, orgasm is not strictly necessary for successful reproduction, and yet it undeniably exists.

Some biologists believe that female orgasm is a critical component of pair-bonding. A sexually satisfied woman, they argue, is more likely to want sex and thus is more likely to hold on to her mate, whose sexual needs are also more likely to be fulfilled. Consider the case of Tom and Jane, two of Judith's patients.

Tom is a handsome surgeon, Ivy League educated, with years of sexual conquests notched on his belt. He was married to Jane, a professional homemaker who had never made love to anyone but him. Tom complained that Jane wasn't sexy enough; Jane wanted a gentle Romeo, more romance, and less huffing and puffing. For ten years the couple had been having frequent but routine sex in a compromise that was minimally satisfactory to both. As long as Tom had enough sex, he stayed in the marriage; as long as Tom paid the bills, Jane provided the sex.

Eventually, Jane wanted out. She made an interesting and calculated choice: to stop the sex. Jane knew Tom well enough to gamble that he would be more generous in a divorce settlement if he left her for another woman than if she left him. She also knew that Tom would not tolerate a life without regular sexual encounters and that he was wealthy, powerful, and attractive enough to get them. So Jane intentionally set her husband up for an affair and then divorced him.

There are many cautionary lessons to be drawn from this couple's story. One is that continuous sexual receptivity is no guarantee of marital bliss. Jane was sexually available to Tom, but she only rarely experienced orgasm. Maybe if she had found sex more satisfying, she would have been a more enthusiastic partner. Unquestionably, a good sexual relationship can contribute to a couple's love. And one way to generate continuous sexual receptivity is to make year-round sex rewarding not only to the man but to the woman as well. Hence the orgasm.

Other theories abound. Donald Symons argues that the female orgasm is an evolutionary vestige, analogous to the persistence of nipples among men. But reproductive physiologists point out that the muscle contractions associated with a woman's orgasm help propel sperm toward its rendezvous with the egg. In his best-selling book *The Naked Ape*, Desmond Morris proposes a behavioral role, suggesting that the female orgasm might help fertilize a woman's eggs by keeping her relaxed and therefore horizontal for a while after coitus. It has also been argued—not very convincingly— that orgasm provides what psychologists call an "intermittent reinforcement schedule" that keeps women interested in and rewarded by sex and, therefore, likely to keep reproducing.

The fact that female orgasm tends to be elusive and difficult to evoke may have an adaptive benefit. Females seeking orgasmic pleasure may be drawn to successful and confident males whose sexual prowess surpasses that of their lower-ranking rivals. Throughout the animal world, subordinate males typically copulate very quickly—if they do so at all— sometimes literally looking over their shoulder lest a dominant male arrive and interrupt them. Dominant males, by contrast, take their time and thus would theoretically be more likely to elicit orgasms from their female partners. It may be telling that in our own species, premature ejaculation is typically associated with young, sexually inexperienced men, those lacking in confidence and self-esteem.

Significantly, women are more likely to experience orgasm in intense, intimate, and loving relationships. After surveying 100,000 of the magazine's unmarried readers, researchers for *Redbook* magazine concluded, "For most women, orgasm depends on being in love and feeling comfortable with their lovers." Women who had sex on a casual basis—once or a few times with several partners—apparently were not doing so for pure sexual pleasure, as they were the least likely to achieve orgasm. Among those who had a series of one-night stands, for example, fully 77 percent said they never reached orgasm. In contrast, women who were having sex in a regular, stable relationship were most likely to be orgasmic. Among this group, only 23 percent reported that they never reached orgasm. The article concluded that "young women need a sustained sense of intimacy, security, and trust from a relationship before they shake off inhibitions and respond sexually."

A similar situation exists for married women. Women who share a close relationship with their husbands in other aspects report high levels of sexual satisfaction. In one study, for example, 64 percent of wives who engaged in extensive nonsexual activities (sports, travel, reading) with their husbands also described being sexually satisfied, compared with only 18 percent of those who did not. Generally, a woman's perception of her marital happiness correlates strongly with the frequency and consistency of her orgasms, which further suggests that orgasm is a way women signal to themselves that they are in a good relationship.

Why do men reach orgasm more quickly than women? A minimalist, evolutionary view of sexual intercourse looks simply at sperm transfer from male to female. This requires nothing more than for the man to ejaculate quickly and for the woman to passively receive his semen. But for women, as for men, orgasm is pleasurable, tension reducing, and positively reinforcing (having experienced one, most people are likely to want more).

Still, the overwhelming majority of female animals, including women, function quite well without orgasm, and there is no evidence that orgasmic women have more offspring. In fact, in some human societies—especially those that practice genital mutilation (clitoridectomy)—female orgasm is apparently unknown. Women who never experience orgasm may be less happily married than they would like but may produce plenty of children nonetheless. By contrast, a man who is nonorgasmic or, more precisely, nonejaculatory, cannot be a biological father.

A woman in the throes of passion may be disappointed by her lover's premature ejaculation, which generally marks the end of intercourse. But from a biological perspective, if sex has to end following orgasm, it should be after the man ejaculates. After all, the woman can stop—and make shopping lists in her head or count the cracks in the ceiling—while the man continues. Once the man stops, however, intercourse usually ends.

Ready and Willing

Women are certainly unusual, and perhaps unique, because unlike other animals, whose sex lives are restricted to a narrow period of time around ovulation, they can mate continuously throughout the entire menstrual cycle. Although some women can detect their own ovulation by the onset of *mittleschmerz*—middle pain—caused by the release of the egg from an enlarged ovary halfway through their menstrual cycle, most require sophisticated techniques to know when they are ovulating. Behavioral changes are subtle as well, with some women reporting heightened sexual interest at this time and others indicating no difference in libido.

In contrast, among other mammals, ovulation is a much-heralded event. Most female mammals advertise when they are in heat and ready, willing, and able to be fertilized. Their behavior changes, they may show physical signs of genital enlargement, and they emit characteristic chemical signals. These cues explain why dogs and other male mammals spend considerable time sniffing the genitalia of females. Stallions not only sniff mares' bottoms but also assume a characteristic facial expression, with wrinkled nose and protruding lip, when they detect estrous odors. Mares in heat present their rumps to be sniffed and then produce a vast amount of strong-smelling urine, which leaves no doubt in anyone's mind (even that of a human observer, who may be comparatively obtuse in such matters) that they are ready to mate. A casual visit to the monkey quarters at a zoo reveals that most sexually receptive monkeys or chimpanzees emit visual as well as chemical and tactile cues. No one—least of all an ardent male chimpanzee—can miss the swollen, red rump of an estrous female.

Why are we humans so secretive, comparatively, about such an important aspect of our own biology? Nancy Burley, a biologist at the University of Illinois, speculates that prehistoric women were probably aware that childbirth could be difficult, painful, and often fatal. According to Burley, women whose ovulation was most conspicuous—at least to the women themselves—might have abstained from sex when they were ovulating and, as a result, would have been less likely to become pregnant. (Of course, this theory assumes that the women could discourage their mates from intercourse at such times.) Who, then, were most likely to become mothers? Women whose ovulation was least apparent to themselves. In other words, women who could detect ovulation might have been intentionally having fewer children while those who could not were unwittingly inheriting the earth.

No doubt our social world would be much different if a woman's ovulatory status were public knowledge. As primatologist Jane Lancaster asked, "What would happen to the division of labor if human females came into estrus? If times are bad and vegetable food scarce, who is going to go hunting if there is an irresistible female in camp?"

In a more modern setting, what would happen at a male-dominated workplace if a female employee were to come to work reeking of genuine sex hormones (and not mere perfume)? Or at a high school filled with young estrous females and eager adolescent males already suffering from testosterone overdose? Even when freshly showered and fully clothed, such women would probably give off powerful signals of sexual receptivity. One can only imagine how much worse sexual harassment might be if a woman's reproductive status and her receptivity to sexual advances were prominently advertised.

Concealed ovulation may offer yet another advantage. If women were sexually receptive for only a day or so each month and broadcast their availability by scent or behavior, men might well become frenzied, if not violent competitors, leaving women with little choice, perhaps, but to accept the victor. According to this line of thinking, by keeping their ovulation secret women would have more choice in mating and would be more likely to remain rational, cool, and in control sexually. To be sure, women are often sexually passionate, but not in the mindless, nymphomaniacal manner of, say, a mare in heat. By foregoing estrus, women remain mistresses of their genetic fate.

For ovulation to be concealed, sexual receptivity must be continuous. And continuous sexual receptivity in turn may be part of what we call the "lion strategy." Few people realize that lions are among the world's sexiest creatures: they copulate upward of 100 times per day for the three or so days when the female is in heat. Some researchers suggest that the female's extraordinary sexual demands make it unlikely that her mate will inseminate other lionesses, thus reducing the likelihood that her cubs will be forced to compete with other litters for food.

By the same token, if a woman keeps a man satisfied—or better yet, exhausted—he won't be inclined, or able, to inseminate anyone else. By concealing ovulation and extending the period when they are sexually receptive, women may increase the chances of keeping their men around. Such a strategy has enormous significance for humans, whose infants remain helpless for a long time and thus need the resources of two parents. If men knew when their mates were fertile, they might stick around for only a few days and then search for other women in estrus, bolstered by the confidence that their wives (now unreceptive and not ovulating) would hold little appeal for other males.

One difficulty with this argument is that many small birds, such as warblers, robins, and sparrows, are monogamous despite having a brief period of sexual receptivity (and, incidentally, no female orgasm). Thus, it is clear that male–female bonding can occur without regular sexual intercourse. But among humans, studies show that sexual relations help provide the glue that keeps husbands and wives together. In fact, surveys of couples who have remained happily married for more than twenty-five years show that sex is an important element in most of these marriages. Certainly, the converse is true: sexless marriages are more likely to end in divorce than are those that are sexually fulfilling.

Change of Life

A final mystery is why women—alone among female mammals—experience a dramatic shutdown in their reproductive systems, typically during their fifth decade of life, when they may yet have thirty to forty years to live. Most living things reproduce throughout their lifetime; when they can no longer breed, they die. In our own species, men follow the same pattern: they produce sperm (although in diminishing numbers and with declining viability) as long as they live. Why are women different?*

It stands to reason that as a woman ages, her chances of being weakened or killed by the rigors of pregnancy, childbirth, and lactation increase. At the same time, by age forty-five to fifty, most women (at least in pre-technological times) will have reared children who are themselves ready to reproduce. At this stage of life, women may contribute more to their long-term biological success by shutting down their own reproductive engines and helping to rear their grandchildren rather than risk giving birth. In fact, a recent study of African hunter-gatherers called the Hadza suggests that this is the case. According to University of Utah anthropologist Kristen Hawkes and her colleagues, grandmothers play a key evolutionary role by ensuring that their chil-

---

*To our knowledge, only one other mammal undergoes true menopause: the short-finned pilot whale.

dren and grandchildren get enough to eat. Dubbed the "grandmother hypothesis," the idea is that grandmothers can help wean their grandchildren from breast milk, thus freeing their daughters to produce more babies. It is irrelevant, by the way, whether modern postmenopausal women function actively as grandmothers or even whether they have grandchildren at all; the point is that they did so for much of human evolutionary history.

What Homosexuality Says About Sex Differences

Why, if evolution favors those who produce the most offspring, does homosexuality exist? After all, individuals attracted to members of the same sex are less likely to reproduce than are those attracted to the opposite sex. Many theories have been proposed; one suggests that nonreproductive homosexuals help rear successful relatives, thus indirectly passing on their genes to the next generation. Another proposes that homosexuality is culturally induced or is simply a conscious choice of lifestyle. Although we are unconvinced by these theories, we believe that there is much that homosexuality can tell us about the biological underpinnings of male–female differences.

To begin with, homosexuality among men would be expected to mirror male sexuality, freed from inhibitions such as monogamy, which are imposed by female sexuality. In other words, if "straight" (exclusively heterosexual) men are prone to the low-parental-investment strategy of being variety seeking, easily stimulated, and sexually pushy, then gay men should be even more so—and they are. Gay men often find their sexual partners by "cruising" well-known bars and—at least before the risk of contracting acquired immune deficiency syndrome, or AIDS, through these activities became widely known—by frequenting the infamous "bathhouse scene" or seeking quick, impersonal, anonymous sex in rest rooms, parks, or other semipublic places. Among gay men, relationships may often begin with sex, unlike heterosexual pairings that are more likely to progress to sex.

Data gathered in San Francisco before the AIDS epidemic show that three-fourths of gay males claimed to have had more than 100 sexual partners and fully one-fourth reported more than 1,000. (Among a comparable sample of female homosexuals, only 2 percent had as many as 100 partners and none claimed 1,000.) Although many gay men have reduced their frequency of new sexual encounters in response to AIDS, they are no less interested in sexual variety. Their behavior demonstrates, rather, that such desires can be overridden by other factors, such as the fear of catching a lethal disease. A gay friend of ours in his fifties who lost his lover to AIDS continues to date and occasionally makes love with new partners. "You just get used to latex," he says, "and playing the odds."

A substantial industry specializes in providing visual stimuli for male homosexuals; nothing comparable exists for their female counterparts. *Playgirl*, originally meant to be a "soft porn" magazine for heterosexual women (displaying unclad men), now appeals more to gay men. Other magazines catering specifically to gay men are characterized by their unrelieved focus on male genitalia, often with minimal or no text.

By contrast, if a magazine ever attempted to reach a lesbian audience by featuring extensive photographs of naked women, it would almost certainly fail—except, perhaps, among heterosexual men. Lesbians simply do not spend time ogling the naked women in *Playboy*, *Penthouse*, and *Hustler*. Further, because they are not turned on by sexual variety per se, lesbians visit lesbian bars more for genuine socializing than in search of sex.

Alice and Kathy have a relationship typical of many lesbian couples. Strictly monogamous, they both enjoy sex but do not find it central to their lives. They frequent a coffeehouse that caters to lesbians, but they go there to drink fruit smoothies and listen to jazz, not to pick up other women. Lovemaking is an occasional thing, on par with attending a good concert: fun, but by their own account, "not a big deal."

In contrast, Alfonso and Scott, a gay couple who have been together for ten years consider sexual fidelity out of the question: both enjoy quick bouts of intense but impersonal sex with strangers. Like other confident male homosexuals, each accepts his partner's desire for sexual variety but within certain guidelines. For Alfonso and Scott, this means that encounters with additional lovers may take place only outside the home. (For other gay couples, additional lovers may be permissible only if brought home and shared, only if talked about, only if not talked about, and so forth.) By contrast, lesbians tend to weave lasting monogamous ties, involving high levels of fidelity.

Donald Symons suggests that the rampant sexuality characteristic of many gay men is not unique to gays. Many straight men, he points out, would be delighted to stop off during their lunch hour for an episode of anonymous fellatio. If casual sex were as socially acceptable, safe, and inexpensive as espresso, we might see sex stations at about the same frequency as coffee shops.

Philip Blumstein and Pepper Schwartz—University of Washington sociologists whose orientation, like that of most sociologists, is distinctly nonbiological—conducted a study that inadvertently confirms much of what we have been saying about homosexual preferences. After interviewing hundreds of American couples (married, unmarried and living together, gay men, and lesbians), they reached the following conclusions:

- "Gay men have sex more often in the early part of their relationship than any other type of couple. But after ten years, they have sex together far less frequently than [heterosexual] married couples. . . . Although interest in sex with their partners declines, interest in sex in general remains high. Sex with other men balances the declining sex with the partner."
- "Most gay men do not care if their partners are monogamous. If a gay man is monogamous, he is such a rare phenomenon, he may have difficulty making himself believed."
- Lesbians "have fewer outside partners than all the other groups, and gay men seek more variety; they seldom stop with a few outside partners."
- When it comes to sex without love, "lesbians are much less in favor of it than are gay men."
- Among homosexual couples, whether male or female, problems arise over who initiates sex. "Many lesbians are not comfortable in the role of sexual aggressor and it is a major reason why they have sex less often than other kinds of couples" whereas among gay male couples, "both feel free to be the initiator [but] having two initiators in a couple can create problems."

Despite these findings, Blumstein and Schwartz adhere to the traditional reasoning of social scientists, that male–female differences in behavior reflect social roles and diverging expectations of what men and women should do. For our part, we do not claim that biology is the sole determinant of male behavior, female behavior, or male–female differences, simply that biology counts. There is a saying that if something looks like a duck, quacks like a duck, and acts like a duck, then it probably is a duck. When male behavior, female behavior, and male–female differences are all fundamentally consistent with biology—and, furthermore, when these patterns hold for homosexuals of both sexes no less than for heterosexuals—then, we believe, it's time to start acknowledging the duck.

Strategies for Choosing a Mate

There has been a good deal of research into what each sex finds attractive about the other, and the results are consistent with biological theory. Men tend to value youth and physical attractiveness as well as the prospect of sexual fidelity (all of which suggest reproductive potential), whereas women value financial assets, status, and signs of ambition and industriousness (all of which contribute to successful offspring). Not surprisingly, peak attractiveness for women correlates with peak reproductive potential. In other words, men are especially likely to find women attractive when they possess traits associated with fertility: youthful appearance, good complexion, healthy-looking hair and the like, not to mention adequate breast and hip development.

When psychologist David Buss, at the University of Michigan, asked students to rank the traits they desired in a prospective spouse, both men and women rated kindness and intelligence as numbers one and two, respectively. After those traits, men listed beauty and youth while women went for wealth and status. Wondering whether this difference in ranking said something about male–female differences generally or just among Americans, Buss then repeated his survey in thirty-seven cultures and found precisely the same result.

Skeptics pointed out that the women in the study may have sought wealth and power in a spouse because this was something they lacked and men had. If women were already wealthy and powerful, they argued, things would be different. So Buss looked harder at wealthy and powerful women. His findings? Wealthy and powerful women sought men who were wealthier and more powerful yet!

Personal advertisements in newspapers provide a fascinating glimpse into the traits that men and women find most appealing. They also reveal predictable differences between men and women, differences that could have been scripted by an evolutionary biologist. Men offer resources and ask for attractiveness and youth, whereas women offer attractiveness and youth and ask for resources.

In a test of how men and women respond to the details of personal ads, women were found to prefer such words as *loving, reliable, monog-*

amous, career oriented, and *emotionally stable*, whereas men responded positively to *good figure, attractive, trim, sexy, good-looking*, and *young*. It may also be noteworthy that longer advertisements held special appeal for women, who, being the choosier sex, can be expected to hold out for more information. In contrast, briefer ads were more attractive to men, consistent with their inclination to home in on sexual availability and attractiveness.

Finally, men on average received fewer responses per advertisement than did women (1.50 to 4.53), precisely what would be expected in a world in which women have something that men want. Although most of this research was conducted in the United States, an almost identical pattern has been reported for personal advertisements appearing in German, Dutch, and Indian newspapers.

Results of a different study emphasize the dichotomy between the sexes regarding what men and women seek in a relationship. In the late 1980s, this lengthy but revealing question was asked of 232 women and 183 men attending college in California: "If the opportunity presented itself to have sexual intercourse with an anonymous member of the opposite sex who was as competent a lover as your partner but no more so, and who was as physically attractive as your partner but no more so, and there was no risk of pregnancy, discovery, or disease, and no chance of forming a more durable relationship, do you think you would do so?"

Fifty percent of the men, and only 17 percent of the women, said yes. The question was then changed, increasing the physical attractiveness of the hypothetical new partner. Is it surprising that men, but not women, became more interested? The question was changed once more to suggest that a long-term relationship might result from the imagined encounter. This time, male interest was not affected but female interest shot up.

Understanding Jealousy

Jealousy, too, leads to predictable differences. Men are more likely to be jealous of women's sexual liaisons, whereas women are especially jealous of men's attentiveness to other women. College couples were asked which situations were especially likely to trigger their jealousy. Sure enough, men were more likely to describe cases in which their partner was sexually involved with someone else and women were more likely to describe their partner spending time with, talking to, or kissing another woman. In a second study, couples were asked to role-play situations that created jealousy. Three women and sixteen men generated scenarios in which the partner was sexually involved with someone else, whereas sixteen women and only four men constructed scenarios in which there was a loss of the other's time and attention.

In yet another revealing piece of research, couples were asked to "imagine your partner trying different sexual positions with another person," and, alternatively, to "imagine your partner falling in love with another person." The majority of men found the first option (sexual infidelity) to be most upsetting, whereas most women found the second prospect (emotional infidelity) more troublesome. These differences also manifested themselves in measurements of heart rate and galvanic skin response ("lie detector tests"). The biological rationale for this difference? Females are especially threatened by the loss of their provider, not by his sexual exploits per se, whereas males are more threatened by uncertain paternity and the possibility that their resources will support another male's offspring.

The onslaught of AIDS, the world's most lethal sexually transmitted disease, has introduced a new threat to relationships: added to the emotional and evolutionary wounds brought about by a partner's infidelity is the very real risk of contracting a deadly illness. One of Judith's patients, for example, sought a divorce after her husband suggested she be tested for AIDS because, as he confessed, he had engaged in more than a dozen affairs during their twelve-year marriage. Although this particular woman was not infected as a result of her husband's transgressions, many thousands of others have not been so fortunate. National news reports recently focused on the revenge killing of a highly promiscuous AIDS-infected man who had knowingly exposed at least twenty women in the Chicago area, some of them as young as sixteen.

Understanding Love

Men and women are not, and never will be, mirror images of each other. Just as Americans and Britons are two people divided by a common language, men and women are two sexes united—and, sometimes, divided—by their sexuality. Men generally want more sex than their fe-

male counterparts, more often, and with more partners and less talk.
Women want longer couplings, fewer partners, and much more talk.
When women say no to a request for sex, it does not necessarily reflect
a lack of love, just a different pattern of libido. When men clamor for
sex, they may not be lacking in love, just in tact. "A woman waits for
me," wrote Walt Whitman. "She contains all, nothing is lacking. Yet all
were lacking, if sex were lacking."

We suspect it was a man who coined the phrase making love as a ro-
manticized shorthand for sexual intercourse because for many men, the
act of sexual intercourse is in fact the primary act of love. Women certainly value intercourse, but for them, "making love" could as easily
mean listening sympathetically or bringing their beloved a latté in the
morning. Women need to acknowledge that to some extent, men really
do see them as sex objects—even in cases of genuine love—and one
does not preclude the other.

Both men and women would also do well to realize that finding an
appropriate mate for a long-term relationship requires thoughtful advertising, careful shopping, and a discriminating outlook. Anxiety over
mate selection and ignorance of biology can lead to poor "shopping"
practices. People in search of mates shop constantly: around the office
water cooler, in the snack bar and the doctor's waiting room, and at
every school, church, grocery store, laundromat, movie theater, rock
concert, art museum, and on and on. The biologically ideal female
strategy calls for cautious advertising and careful comparison shopping
until a mate is found with the right combination of genes, behavior, and
resources. Women need to be aware of the cost of reproduction and the
difficulties of maintaining male investment in offspring.

It is harder to delineate an ideal male strategy. Promiscuity may result in a lot of offspring, but without attention they may not prosper.
Monogamy, however, may leave a man frustrated, wondering what sex
with many different women would be like and desiring others. In a society in which monogamy is valued and both males and females are expected to provide parental investment, it pays for both men and women
to be careful comparison shoppers as well as good advertisers.

Men ought to keep in mind that women are generally attracted to
those who do well at work (and thus can be counted on to provide resources), who appear physically healthy and robust, who express genuine interest in them as people, and who can be trusted to be faithful.

Women need to compute carefully the costs of a potential mating, including the risks of pregnancy, deception, and abandonment. Virtually any woman can achieve copulations; the challenge is to achieve a bonded relationship with the right partner.

Humanity's long polygynous history helps explain some of the more troubling male–female differences in sexual inclinations—for example, why men typically seem to commit adultery more often than women. This is not to condone adultery, although an understanding of its root causes may help couples in their search for a fulfilling sexual relationship within monogamous marriages. We have yet to see an "open marriage" that really works. More often, avowedly consensual open marriages are smoke screens for one party to indulge in a promiscuous lifestyle, while the other grudgingly consents, rather than leave altogether. Famous threesomes, such as Henry and June Miller and Anaïs Nin, are also infamous for their fireworks.

In her "General Review of the Sex Situation," Dorothy Parker neatly summarized some of the differences between men and women:

> Woman wants monogamy;
> Man delights in novelty.
> Love is woman's moon and sun;
> Man has other forms of fun.
> Woman lives but in her lord;
> Count to ten, and man is bored.
> With this the gist and sum of it,
> What earthly good can come of it?

Frankly, we don't know the answer. But we are convinced that a greater understanding of how biology influences sex differences can lead to greater sensitivity, tolerance, appreciation, and even love.

CHAPTER 4

Violence

HE WHO HAS NEVER struggled with his fellow-creatures is a
stranger to half the sentiments of mankind.
— Adam Ferguson, "An Essay on the
History of Civil Society," 1767

Until a decade or two ago, it appeared that no other
animal, apart from humans, killed members of its
own species. But long-term field studies in animal
behavior have since dispelled this myth. Chimpanzees kill others of their own kind, as do wolves, lions, elk, and bison. In
fact, nearly every mammal species that has been carefully studied has sooner or later revealed a penchant for lethal violence.
Biologists also know that when such events occur, the perpetrators are almost always male.

Among human beings, biology has set the stage not only for
the fabled "battle of the sexes" but also for battles *within* the

sexes, especially those pitting males against males. As with other animals, violence among humans is by and large something men direct toward other men; they are disproportionately both the perpetrators and the victims.

Men, in fact, are so much more violent and deadly than women that the difference is taken for granted. On hearing the term *suicide bomber* or *serial killer*, most people automatically—and correctly—assume that the individual in question is male. Across the board, the human "killing establishment"—soldiers, executioners, hunters, violent gangs, even slaughterhouse operators—is overwhelmingly male. From rampages in post offices and the infamous California schoolyard massacre to unprovoked, deadly shootings in a Texas restaurant, the Long Island Rail Road, and the Empire State Building, men—not women—are the mass murderers. Nor is this imbalance limited to the United States. When killings take place, whether in Bosnia, Rwanda, Cambodia, El Salvador, or Israel, the culprits are nearly always men.

The same applies to the uncountable private episodes of violence that receive little national attention but are the stuff of many a personal tragedy. Admittedly, an occasional Lizzie Borden or Lorena Bobbitt surfaces, but for every Bonnie, there are about a hundred Clydes. In fact, male brutalizers and killers are so common that they barely make the local news. Female killers, however, always achieve a kind of fame; for example, when Susan Smith drowned her two sons in 1994, she received international attention. Yet when men kill their children, they get comparatively little notice. Although they are no less tragic, such events are simply too commonplace to generate more than local dismay.

For a murderous man to generate a response comparable to that sparked by a woman who kills, his crime must be especially dreadful, as in the case of serial murderer Ted Bundy, who not only killed young women but was charming to boot, or cannibal Jeffrey Dahmer, or he must be a celebrity, like accused killer O. J. Simpson. Violence may or may not be as American as cherry pie, but it is certainly a male proclivity. It would not be realistic to romanticize or idealize women or to deny that they too can sometimes be violent and deadly, but when it comes to brutal behavior, the two sexes simply are not in the same league. Why?

Before answering this question, we must emphatically state that by looking to evolution to explain the violence and conflict that rocks our society, we are in no way attempting to justify or legitimize such behavior. On the contrary, we hope that by bringing to the fore the evolutionary logic that is inherent in human conflict worldwide, that we may help bring about a better understanding of our species and so ultimately *reduce* the toll of human violence. To deny the connection, we think, is akin to putting one's head in the sand. The reality is that all social animals—whether parrots, peacocks, or people—engage in conflict. Add sex differences to the equation and the stage is set for even more trouble.

Advocates of social learning theory point out that men are expected to be aggressive and women are supposed to be more passive. Thus, they claim, people grow up meeting the expectations that society imposes on them. One such advocate of this theory, British psychologist Anne Campbell, thinks that men are more aggressive than women because men and women interpret their aggressive tendencies differently: Women see aggression as a loss of self-control and are ashamed of their anger, associating it with antisocial behavior. In contrast, men view their aggression positively; for them, it is a way to *gain* control. Campbell's analysis is probably correct so far as it goes. But it doesn't go far enough. Why, for example, do men view controlling others as more important than controlling themselves? If the answer reflects societal influences, why are identical sex differences found in just about every culture on earth? And why do similar patterns exist in other species?

Why Men Are More Violent

The difference in reproductive strategies between males and females— with males varying greatly in the number of offspring they produce and females varying not much at all—holds the key to patterns of violence. In a nutshell, males must compete for access to females either through song, coloration, or display or by engaging in direct battle with their competitors, and thus evolution has strongly favored aggression over timidity. Cross-culturally, aggressiveness is widely—and all too correctly—seen as manly, and its opposite, timidity, is seen as womanly. (A statement by President Lyndon Johnson provides a memorable example of this. When told that a high-ranking member of his administration had become a dove on Vietnam, Johnson snarled, "Hell, he has to squat to take a piss.")

Levels of aggressiveness correlate nicely with mating strategies.

Among monogamous nonhuman species, such as geese, eagles, foxes, gibbons, and most songbirds, males and females produce nearly equal numbers of offspring and also are nearly equal in physical size and aggressiveness. Among polygynous species, however, the bigger and nastier a male is, the more likely he will be to fend off his competitors successfully and win the mating game. Accordingly, it is the James Bonds and the Rambos, not their more pacific brothers, whose genes are projected into the future, thus giving rise to succeeding generations that are likely to be, if anything, more violent.

When does this arms race stop? Only when the overall disadvantages of such behavior exceed its evolutionary benefits. At some point, highly aggressive individuals either run too great a risk of injury or death or lose out in other ways. For example, among some birds, males occasionally spend so much time singing, posturing, threatening, and fighting with their male neighbors that they neglect their own offspring. Overall, however, natural selection smiles on behavior—any behavior—that contributes to reproductive success. For males, that smile has been especially broad and toothsome when it comes to aggressiveness.

This link between gender and aggression is clearly seen in the strange case of the blue-headed wrasse, a polygynous species of coral reef fish. Breeding groups consist of a relatively large male associated with a bevy of females. But there are no bachelor males skulking resentfully in the coral crevices, as there are, for example, unsuccessful harem master wannabes among elephant seals. Instead, blue-headed wrasse populations start off as all female. On maturity, the largest and most aggressive individuals become male—in this species, sex is not determined by the X or Y chromosome but by hormonal changes triggered by behavioral events. Remove the male from a breeding group and the largest, most aggressive female will stop producing eggs and turn into a sperm-producing male. Within a week, egg-making ovaries become sperm-making testes, and the newly transformed male mates with his bevy of females. For our purposes, this bizarre example of transsexualism demonstrates, almost diagramatically, the essence of the relationship between maleness, aggressiveness, and biology. A blue-headed wrasse is biologically rewarded for being male *if* it is large enough and tough enough to dominate the others. Otherwise, it remains meek and female.

Ethologists, who spend their careers studying animal behavior primarily under natural conditions, have known for a long time that in most species, male–male aggression is far more frequent and violent than female–female aggression. For example, whereas female cats, whether the household variety or African lions, may snarl and hiss, posture, and bat at one another, male cats are the ones that kill their fellows. Among chimpanzees, the frequency of female–female aggression is only about one-twentieth of its male–male counterpart.

Primatologist Franz de Waal of Emory University describes "aggressive politicking" among male chimpanzees, which form potentially violent coalitions according to shifting rivalries and incentives of threat and reward. Male rivals often meet an untimely death at the hand of such coalitions. In contrast, coalitions of female chimpanzees are oriented toward supportive family relationships rather than murderous competition.

Why has evolution favored such a distinct gender gap? Simply put, males succeed reproductively at the expense of fellow males, whereas a female's reproductive success is unlikely to be enhanced by knocking fellow females out of the way. If anything, females with a penchant for ferocity are more likely to suffer injury with little or no reproductive gain to show for their efforts.

On occasion, of course, females can be violently competitive. Dominant female African hunting dogs may kill the offspring of lower-ranking females; female red howler monkeys push around other females; and female groove-billed anis (ravenlike tropical birds) sometimes evict rivals' eggs from their communal nest. Female–female competition is undoubtedly more widespread than many people realize, but in almost all cases it is less direct, less boisterous, and certainly less violent than its male–male counterpart.

Primate specialist Sarah Hrdy agrees that female–female competition among humans often goes unnoticed:

> Consider . . . such phenomena as sisters-in-law vying for a family inheritance which is to be passed on to their respective children, or the competition for status between mothers. . . . The quantitative study of such behavior in a natural setting hardly exists. We are not yet equipped to measure the elaborations upon old themes that our fabulously inventive, and devious, species creates daily. . . . How do you attach a number to calumny? How do you measure a sweetly worded put-down?

Such catty competitiveness may arise before puberty. On several occasions, Judith has been asked to intervene on behalf of girls who developed anxiety and social phobias as a result of being traumatized at school. The victims were socially ostracized by their peers and made the butt of telephone whispering campaigns, which occasionally escalated to the point of name-calling. Such tactics are exceedingly rare among boys, who are much more likely to use direct intimidation or physical violence.

The pattern continues among adults. Thus occasionally women kill their husbands, their ex-husbands, or the wives of their lovers, but the few who do are considered endlessly fascinating. Jean Harris, headmistress of a tony girls' boarding school, became infamous when she murdered Dr. Herman Tarnower, author of *The Scarsdale Diet*, after finding his new lover's nightgown, as did Joey Buttafuoco's teenage lover, Amy Fisher, when she put a bullet through his wife's head.

But overall, such acts fail to hold a candle to the brutal, bloody violence that so often characterizes competition among men. War making, like murder and other forms of violence, is almost entirely a man's activity. Even in the United States today, only a handful of women engage in military combat. There has never been a society on earth in which women exceeded men as war makers; in fact, they have never even come close.

Nor has a single woman perpetrated widespread genocide on the scale of Hitler, Stalin, Pol Pot, Idi Amin, Tamerlane, Genghis Khan, Caligula, and the like. Indeed, one might be hard-pressed during a game of Trivial Pursuit to name several bloodthirsty women in all of history, whereas the list of male contenders seems endless.

Admittedly, men are overwhelmingly in political and military control, but even so, the male inclination to be soldiers and warmongers is tightly linked to biology. If violent behavior reflected the vagaries of culture, we would expect female-initiated violence to be as frequent and intense as its male counterpart; it assuredly is not. The evidence is overwhelming that men and women are fundamentally, biologically different.

Studies of prosecution and imprisonment records in Europe going back several centuries, as well as examinations of modern crime statistics from the United States and around the world, show that men consistently outstrip women when it comes to violent crime by a ratio of at least three or four to one. The same applies to crimes against property. Men (especially those who are impoverished) are far more likely than women to rob their victims face to face, in a manner reminiscent of dominant–subordinate upmanship. The only areas in which women commit more crimes than men are prostitution (which some say is not a criminal activity but an act between consenting adults) and shoplifting, which is nonconfrontational.

When women are aggressive, their behavior tends to be defensive, as when a woman kills a man who abuses her or her children or when she "fights" to have her child's murderer condemned to death. Among animals, a mother bear with cubs, for example, is notoriously fierce, as are other females when defending their young. Thus, women's aggression tends to be reactive, whereas that of men tends to be truly "offensive."

Another interesting fact is that the male sex hormone testosterone is associated with violent crime among men. The higher the level of testosterone in a man's bloodstream, the greater is his tendency to be aggressive, even ferocious.* Interestingly, high testosterone levels are also correlated with violent crime among women. When researchers looked closely at testosterone levels in eighty-four female prison inmates, they found that those convicted of unprovoked violence had the highest testosterone levels, whereas those incarcerated for defensive violence (who had responded in self-defense after being physically assaulted) had the lowest levels. Testosterone levels also tended to be high in women who had long criminal records or who had been declared dangerous by their parole boards. Put another way, women carrying a "macho" dose of hormones are likely to behave like men.

To make matters worse, males are to some extent caught in a vicious circle: their penchant for violence makes them vulnerable to get more violence. Thus the male proclivity for risk taking and competition often results in accidents, fights, and drug and alcohol binges. Yet when a person's brain is injured, as by physical damage or the effects of drugs or alcohol, his or her behavior becomes less thoughtful and more vio-

---

*In 1996, a report was released showing that administration of testosterone can actually *reduce* feelings of aggressiveness and unease among men. In the media hullabaloo that followed, few people noticed that the research dealt with men whose testosterone levels were pathologically low; administering additional hormone to them simply reestablished normal levels. It is not surprising that increased amounts of testosterone help reestablish a degree of calmness not previously present. But these findings in no way demonstrate that testosterone is a "calmness" hormone!

lent. There even exists a term, *dementia pugilistica*, that refers to brain damage from repeated blows to the head. Ironically, then, not only are males more aggressive than females, but their aggressiveness also makes them more likely to sustain brain injury, which in turn further increases their chances of resorting to violence.

Although the precise mechanism remains obscure, men are more known to be vulnerable than women to mental illnesses that produce violence. For example, adolescent boys suffer more than do adolescent girls from what is descriptively labeled "oppositional defiant disorder" as well as "general conduct disorder." And in a wide range of impulse control disorders, including intermittent explosive disorder, pathological gambling, and pyromania, males greatly outnumber females. It is noteworthy that men who lack impulse control are likely to be violent, whereas women tend toward kleptomania or trichotillomania—that is, they shoplift or pull out their hair.

Getting Away with Murder

Hit, dust, waste, take down, pop, rub out, off: all are euphemisms for murder. When it comes to homicide, again men are far and away the most frequent perpetrators, as would be expected of a behavior that has its roots in male–male competition.

After reviewing murder records over a wide historical range and from around the world, psychology professors Martin Daly and Margo Wilson of McMaster University in Ontario, Canada, concluded "There is no known human society in which the level of lethal violence among women even begins to approach that among men." More specifically, they found that a man is about twenty times more likely to be killed by another man than a woman is by another woman. This finding holds true for societies as different from one another as modern-day urban America (Philadelphia, Detroit, and Chicago), rural Brazil, and traditional villages in India, Zaire (now the Democratic Republic of Congo), and Uganda. This is not to say that murder rates are equivalent in these places. In modern Iceland, for example, 0.5 homicides occur per 1 million people per year, whereas in most of Europe the figure rises to 10 murders per million per year, and in the United States it soars to more than 100. In all cases, however, male–male homicide exceeds its female–female counterpart by a whopping margin. The fact that the pattern of violence remains remarkably consistent from place to place and parallels male–male competition seen in other species argues forcefully for its biological underpinnings.

During 1995 in the United States, for example, 3,329 men were convicted of murder, compared with 226 women. What's more, the victims were predominantly men: 3,051 men versus 508 women, numbers that clearly show men's tendency to kill other men. Moreover, around the world and throughout history, the age of most male murderers (that is to say, most murderers) has remained remarkably constant, in the early twenties. Put another way, those most likely to kill are men at their physical peak who are trying to establish themselves socially and reproductively. Today's proliferation of guns has changed these statistics, but not dramatically. In the United States, for example, the age group with the highest arrest rate for murder is currently those from eighteen to twenty-two.

Judith is familiar with the case of "Big X." By age nineteen, Big X had been arrested twice for drug possession and—by his own admission—had been involved in two armed robberies and a rape. When asked about his life on the streets, Big X replied, "It's pretty good bein' bad." When asked to elaborate, he explained: "The biggest, baddest dudes get the best stuff. You know: respect, clothes, whatever junk you want, and the best chicks." Asked about the chicks, he said, "There ain't a lot of chicks in the gang, but you know, they sure ain't goin' down for the guys at the bottom."

After Big X worked his way to the top of his gang, he was confronted by "Rutter," an imposing kid who had moved into the neighborhood and asked to join. Eventually Rutter was allowed to "jump in" (join the gang), but only after getting beat up as a test of his toughness. From the start, Big X didn't like this rival male and warned him, "You touch my chick, I'll bust your dick." As it turned out, Rutter never touched Big X's girlfriend, but he did look at her and make a provocative comment about her breasts. In response, Big X calmly pulled out a 9-mm pistol and shot Rutter twice—in the groin. Rutter survived, and Big X is now serving a thirty-year term for reckless endangerment and assault with intent to kill.

Other Societies, Similar Patterns

Among traditional peoples, men who compete successfully with other men mate more often and have more children than do their lesser rivals. An early study by Northwestern University anthropologist William Irons showed that among the Yomut Turkmen of Iran, cultural success was rewarded by biological success: wealthier men had substantially more offspring than those who were poorer. Similar correlations have been found virtually everywhere they have been sought.

Indeed, when psychologist Laura Betzig of the University of Michigan looked at a historical cross section of 104 human societies, she found that "in almost every case, power predicts the size of a man's harem." Minor kings would typically have a harem of about 100; kings of greater substance, perhaps 1,000; and emperors, 5,000 or more. Betzig also found, significantly, that dominance is a powerful predictor of harem size.

At the same time, it is fairly obvious that rich and successful people today do not necessarily have more children. This fact isn't altogether surprising: in modern society, means and ends of reproductive success have become disconnected. Yet evolutionary echoes linger on. In a study of French Canadian men, for example, no connection was found between socioeconomic status and reproductive success. However, when the researcher probed deeper and considered number of copulations as well as number of sexual partners, it was possible for him to estimate "number of potential conceptions" had birth control not been used. The results showed that without contraception (the situation throughout most of human history), today's wealthier, more successful men would in fact be producing many more children than would poorer men.

Some of the most pathbreaking and rigorous studies of violence among small-scale, traditional cultures have been conducted by anthropologist Napoleon Chagnon of the University of California at Santa Barbara. Since the late 1960s, Chagnon has periodically lived with the Yanomamö Indians of Brazil and Venezuela. Inhabitants of the rain forest, they call themselves the "fierce people," and for good reason. Within their own villages, Yanomamö men are very pugnacious, regularly engaging in social interactions that involve a lot of bluff and bluster and no small amount of violence as well. Most disputes (which break out frequently) take place over women and are settled by chest-pounding duels or club fights in which the contestants take turns smashing each other on the head. Men strut about seeking to establish their reputations as warriors. Realizing the odds, they memorize defiant speeches to be uttered if they are mortally wounded. According to Chagnon, 44 percent of all Yanomamö men aged twenty-five years or more have killed someone, and fully 30 percent of all adult male deaths result from such violence. In addition to fighting among themselves, men in a typical Yanomamö village devote considerable time and energy to making war on their neighbors. Once such disputes are started, there is an unending cycle of retribution, with a victim's relatives retaliating against the killers or at least against the killer's village or kin. A failure to retaliate would label them as weak, easy marks, and thus vulnerable to further attacks. Not surprisingly, nearly 70 percent of all Yanomamö adults have lost a close relative to violence.

Chagnon concludes that there is a clear evolutionary payoff to this male-generated violence: men who have killed have more wives and more children than do men who have not. One renowned fellow named Shinbone had 11 wives, 43 children, 231 grandchildren, and—at last count, in the early 1980s—480 great-grandchildren. We don't know how many men Shinbone killed, but we are confident that he wasn't meek and mild mannered.

When Chagnon commented to his Yanomamö friends that some anthropologists believed the Yanomamö fought over food—especially animal protein—they laughed and responded, "Even though we enjoy eating meat, we like women a whole lot more!"

Violence at the Bottom

Male–male competition doesn't always afflict the winners. Men can be as ferocious when trying to avoid the bottom of the sociosexual hierarchy as when trying to rise to the top. In fact, battles at the lower end of the competitive ladder are often more vicious than those among the elite. This is probably because men at the bottom have little to lose and thus are drawn to no-holds-barred fighting, a last-ditch bravado involving risky and deadly tactics.

Data gathered in the United States confirm this notion of violence at the bottom. Across the board, killers are more likely to be unmarried, unemployed, less educated, and of lower socioeconomic status than nonkillers. In addition, young men, especially those from disadvantaged social and ethnic groups, are overrepresented when it comes to drug addiction, violent crime, absentee fatherhood, and the like.

The proliferation of violent gangs speaks to the desperation of the have-nots. A young man must prove he is tough enough to fight his rivals and willing to defend his gang at all times. Thus, gang members engage in an endless series of offensive attacks and retaliation, battling those who wrong them or get in their way. It is not unusual in some inner-city neighborhoods to see guns brandished from car windows as gang members careen through the streets displaying their bravado and willingness to fight.

One teenager arrested recently for attempted murder said that his victim looked at him the wrong way. When this offense took place, he and his attacker were separated by a busy street, so nothing happened. But they met the next day when the attacker happened to be cruising by in a car. He pulled over, jumped out, and pumped five bullets into his victim. His explanation for such cold-bloodedness? "I'm the toughest guy on the block."

As with the Yanomamö, retreating from or avoiding violent confrontation brands one a sissy, a loser. Naturalist and explorer Peter Matthiessen notes that among the Dani people of the New Guinea highlands,

> A man without valor is *kepu*—a worthless man, a man-who-has-not-killed. The kepu men go to the war field with the rest, but they remain well to the rear. . . . Unless they have strong friends or family, any wives or pigs they may obtain will be taken from them by other men, in the confidence that they will not resist; few *kepu* men have more than a single wife, and many of them have none.

Manuel Sanchez, a thirty-two-year-old man from Mexico City, sums up the situation nicely:

> Mexicans, and I think everyone in the world, admire the person "with balls," as we say. The character who throws punches and kicks, without stopping to think, is the one who comes out on top. The one who has guts enough to stand up against an older, stronger guy is more respected. If someone shouts, you've got to shout louder. If any so-and-so comes to me and says, "Fuck your mother," I answer, "Fuck your mother a thousand times." And if he gives one step forward and I take one step back, I lose prestige.

But if I go forward too, and pile on and make a fool out of him, then the others will treat me with respect. In a fight, I would never give up or say, "Enough," even though the other was killing me. I would try to go to my death, smiling. That is what we mean by being *macho*, by being manly.

The pattern begins early in life. "Boys will be boys" is the indulgent observation of many, especially those from an older generation, when a boy behaves aggressively. Famed evolutionary biologist and Harvard University professor Edward O. Wilson reflects on his own childhood:

> My worst difficulties came from the fist fights. They were merciless and brutal. . . . One boy, usually the local bully or the "champion" of a group, challenged another boy, usually the newcomer. . . . It was unmanly to refuse a fight. . . . My face was sometimes a bloody mess; I still carry old lip and brow split scars, like a used-up club fighter. Even my father, proud that I was acting "like a little man," seemed taken aback.

Another aspect of male violence is the ease with which it is triggered. After interviewing convicted killers in Philadelphia, sociologist Marvin Wolfgang identified twelve categories of motive. Far and away the largest, accounting for fully 37 percent of all murders, was what he designated "altercation of relatively trivial origin; insult, curse, jostling, etc." In such cases, people got into an argument over something as unimportant as a sports game, who paid for a drink, an offhand remark, or a casual insult. A friend of ours who is a public defender tells the story of a murder that took place in St. Paul, Minnesota. In this instance, a nineteen-year-old boy, who was known to have a quick temper, shot and killed his fifteen-year-old brother. After the two had argued over who should play Nintendo first, the elder brother went into his bedroom, loaded his gun, came back, and shot the younger brother.

To die over something so inconsequential as a casual comment or a dispute about some distant event or ill-chosen word seems the height of irony and caprice. But in a sense, disputes of this sort are not trivial, for they reflect our evolutionary past, when personal altercations were the stuff on which prestige and social success (and ultimately biological success) were based. In this context, it is very upsetting to be "dissed." Thus, it is not surprising that young men today fight and die over who said what to whom, whose prestige has been challenged, or whose clothing is offensive.

Sex and Violence

One of the truisms of ethology has long been that fear inhibits male sexual behavior, whereas aggression and sex are intricately linked. A self-conscious or fearful male may have trouble attaining an erection, and in fact, among human beings, both erectile disorders and premature ejaculation are associated with performance anxiety—in other words, with fear. In contrast, among many species, a sexually aroused male is also aggressive and vice versa. Moreover, aggression often reaps sexual success. For example, dominant squirrel monkeys indicate their social status by displaying an erect penis.

Laboratory studies provide some of the most compelling evidence for the link between male sex and aggression. One clue comes from the simple, naturally occurring chemical, nitric oxide, which has a dampening effect on both. When laboratory mice are prevented from synthesizing nitric oxide, they not only become exceptionally aggressive toward other males but also exceptionally persistent in their sexual advances toward females, continuing both their aggressive and sexual attacks far beyond the limits seen in normal animals.

Sex and aggression are even linked anatomically in males. Neurobiologist Paul MacLean of the National Institutes of Health reported that electrical stimulation of a brain region known as the limbic system causes a squirrel monkey to go from indifference to sexual stimulation, to placidity with penile erection, to intense sexual fervor, to signs of rage (teeth bared, loud vocalization, hair on end), all within an area of the brain no bigger than a square millimeter, or about the size of a pinhead.

Yet another indicator of the link between sex and aggression among males is the worldwide prevalence of phallic symbolism, which in all cultures signals power, dominance, and threat. Slang terms such as "Up yours," "Screw you," and the like, have their equivalent in virtually every language and are always aggressive and demeaning. However we may value sexual pleasure, no one looks forward to being "screwed," "shafted," or made out in other ways to be the victim of sexual abuse.

But the bottom line is that aggression often reaps sexual success . . . if, that is, the aggressor is male.

The link between sexuality—especially male sexuality—and aggression is underscored by cases of homosexual rape in prisons. According to a study of sexual assaults among prisoners, the majority of male rapists deny that they performed a homosexual act. The rape of another man while under incarceration is far more an act of sexual aggression and of male–male competition than it is an expression of homosexuality.*

Rape

In contrast, heterosexual rape seems to be fundamentally different, an unconscious reproductive strategy enacted by males who have an otherwise low probability of reproductive success. It is distressingly common among human beings . . . and among a wide variety of other species as well.

Most animals—like most humans—go through a period of courtship before mating. The two participants may bow, sing, prance, and strut, bill and coo in romantic synchrony, or otherwise follow an elaborately choreographed and predictable pattern that eventually results in their becoming consensual sexual partners. Although copulation among animals may not meet the human definition of "romantic," it is at least likely to be well synchronized, smoothly accomplished and mutually arrived at. Clearly, most cases of sexual intercourse among animals are not rape.

But sometimes things go awry. Male mallard ducks, for example, engage in an especially brutal pattern of forced copulation. The act begins most commonly when the drake is some distance away, and it unfolds much as might a human gang rape. A small flock of unmated males swoops down on a hapless female, which struggles vigorously, trying to escape. Absent are the shared niceties that typify harmonious courtship between a pair of mated mallards. Females sometimes drown in the

---

*A kind of rape occurs among female prisoners as well, but the frequency and degree of violence and subjugation of such acts are far lower than among men.

process, but enough survive to bear their victimizers' offspring, thus perpetuating their genes.

Rape has been similarly documented among fruit flies, mole crabs, scorpion flies, crickets, desert pupfish, guppies, blue-headed wrasse, bank swallows, snow geese, other species of ducks, African bee-eaters, laughing gulls, tree shrews, elephant seals, right whales, bighorn sheep, and wild dogs. It has also been reported among such primates as rhesus monkeys, talapoin monkeys, vervet monkeys, stump-tailed macaques, Japanese macaques, spider monkeys, gray langurs, gorillas, chimpanzees, and orangutans.

According to the National Women's Study, which was released in 1992, 13 percent of adult American women have been raped at least once, 75 percent of them by someone they knew. (Rape statistics typically underestimate the frequency of the crime because many victims, for various reasons, fail to report the incident. Add the number of husbands who force themselves on their wives but are never reported and the number of rape victims rises significantly.) Regardless of the exact percentage, the pattern is clear: with distressing frequency, men—like other male animals—use force to achieve sexual relations.

Many sociologists and psychologists see rape as simply a crime of violence against women. For example, Susan Brownmiller, author of the best-selling *Against Our Will*, said rape is "a conscious process of intimidation by which all men keep all women in a state of fear." Another expert writes: "In terms of the perpetrator's motives, rape bears a closer resemblance to violent crimes such as assault and robbery than it does to sexual intercourse with a consenting woman." To some extent, they are right. Certainly the sexual component of rape has nothing in common with the caring and sharing associated with normal, healthy sexual relations. But a sexual component is nonetheless there, making rape as much a sexual crime as it is a violent one.

Revelations about sexual abuse and rape in the United States military have focused attention on the dangerous consequences of placing some men in positions of strict disciplinary authority over young women. But it would be a mistake to attribute such outrages as the rape of enlisted women by their drill sergeants to disparities in power alone.

The Crime Victims Research and Treatment Center at the Medical University of South Carolina defines human rape as "an event that occurs without the woman's consent, involves the use of force or threat of force and involves sexual penetration of the victim's vagina, mouth, or rectum." Forced penile–vaginal intercourse is not only the most psychologically troublesome form of assault for both the victim and her family but also the most common. The simple and depressing fact is that rapists—whether mallards or men—experience sexual excitement, culminating in most cases in ejaculation and in some cases in pregnancy.

The fact that rape is widespread and is in a direct sense biological in no way makes it normal or acceptable. Clearly, rape involves abuse of power as well as sadism and anger. In fact, in response to atrocities committed in Bosnia, where rape was used as a means of social subjugation and a form of "ethnic cleansing," it has been declared an international war crime.

With a clear understanding that rape is a wholly unacceptable act of aggression, viewed from a strictly biological perspective it can also be seen as an alternative reproductive strategy, typically perpetrated by men who are incapable of achieving or maintaining mutually loving relations with women. Sadly, rape may be their only shot at reproduction.

Statistics bear out the biological basis for rape. To begin with, a review of police records shows that most rape victims are women of peak reproductive age. Young girls (less than ten years of age) and older women (over forty) are less likely to be raped than would be expected from their proportion in the population. True, eighty-year-old women and infants are sometimes raped, but such incidents are perpetrated by men who suffer from an even higher level of pathology than the typical rapist.

In addition, women who are timid or submissive are more likely to be targeted than are women who are domineering or powerful, the opposite of what might be expected if rape were strictly an expression of male anger. Furthermore, if rape were just another violent crime, like assault or murder, rape victims should parallel the age distribution of women who are victims of these other violent crimes. Not so: victims of rape are generally younger than, for example, murder victims.

Furthermore, an evolutionary theory of rape predicts that the rapists themselves would be disproportionately young, and they are. Like young males of other polygynous species that are just entering the breeding population, young men are especially prone to violent, high-risk strategies. Indeed, the stereotypical rapist is, on average, single, socially and economically disadvantaged, and a participant in what Marvin Wolfgang calls the "subculture of violence." In this limited regard, rapists bear a similarity to bachelor elephant seals. They are the dregs of society, exploiting avenues of last resort. Significantly, most rapists are twenty-five to forty years old, at an age when their sex drive remains high, but bitter experience has already shown them that they are unlikely to succeed in achieving a healthy man–woman relationship. One might conclude, therefore, that rape carries "at its diseased heart, a small pressure of genetic advantage . . . being . . . among other things, an automatic, unconscious reproductive strategy of low dominance males" who cannot expect to win a female's affection through means of conventional courtship.

Accordingly, under an evolutionary theory of rape, perpetrators might be expected to be more prevalent among men from lower socioeconomic classes, and they are. In the United States, the frequency of rape is consistently higher among African Americans, who as a group are lower on the socioeconomic ladder, than among Caucasians. Some of the discrepancy may reflect a greater societal willingness to convict minorities of such crimes, as well as the greater ability of wealthy Caucasians to manipulate the legal system. But it probably does not explain the twelvefold difference commonly reported. A more likely explanation is that high-status men of any race or ethnicity have greater access to women and thus have less need to resort to violence to get what they want. Let us be perfectly clear on this matter: race as such is not the issue. If African Americans or, indeed, members of any racial group are prone to rape, it is strictly because members of any group that is socially and economically disadvantaged are vulnerable to such behavior, not because of skin color or any other racial attribute as such.

Crime statistics support the socioeconomic correlation. For example, data from the 1960s show that women living in inner-city areas had a 1 in 77 chance of being raped in a given year, whereas in more affluent areas the risk declined to 1 in 2,000. In the wealthiest suburbs it plummeted further, to 1 in 10,000. The pattern holds for other societies as well: in Denmark, for example, 59 percent of rapists are unskilled workers. Even in places as diverse as central India and Lusaka, the capital of Gambia, women are especially fearful of being raped by poor men.

Such findings do not suggest that all rapists are social failures, or that all social failures are rapists. Certainly, successful upper-class men rape and plenty of unsuccessful men do not. However, we believe that the tendency to rape persists because the gratifications reverberate with long-ago, last-ditch strategies on the part of males who were otherwise doomed to evolutionary failure.

Biologist Randy Thornhill and anthropologist Nancy Thornhill, formerly of the University of New Mexico (from whose work much of our discussion of rape is derived), also made some interesting generalizations about rape. They postulated that rape would be especially frequent in societies where a prospective husband must buy his bride—that is, pay the bride's family for the privilege of marrying her—and they found exactly such a correlation among the Gusii people of Kenya. According to anthropological data collected during the 1930s, Gusii brides were expensive, costing a man eight to twelve cows, one to three bulls, and eight to twelve goats. As a result, many younger men couldn't afford a wife, and both livestock theft and rape were commonplace. During the years when marriage prices went down, so did the crime rate. When the bride price climbed again, up went the frequency of theft and of rape.

Thornhill and Thornhill concluded that worldwide, rapists have not only poor economic and social resources but also low self-esteem: "There are certain general characteristics that men who are prone to rape appear to have in common. . . . His life appears to hold little pleasure and to offer few rewards. His overall mood state . . . is . . . characterized by dull depression, underlying feelings of fear and uncertainty and an overwhelming sense of purposelessness and hopelessness. At the root of this are deep-seated doubts about his adequacy and competency as a person."

Thornhill and Thornhill also noted that being raped adversely affects a woman's relations with her husband or boyfriend. This is because of the possible reproductive consequences of being raped; the emotional impact is not necessarily lessened even if a husband knows that his wife has not become pregnant. In addition, they found that the lower the social standing was of the offender, the more the victim was traumatized by the event, assuming that other factors, such as degree of violence, length of sexual intercourse, and absence of disease were equal. Furthermore, rape involving penetration was the most emotionally traumatic. Not uncommonly, rape victims in some cultures are cast out by their families, a response that only adds to their terrible emotional burden.

Judith, for example, treated a young Thai woman who had been raped and knifed. While the woman was in the hospital for treatment of a severe liver laceration sustained during the rape, her family disowned and disinherited her because they believed she had been defiled and was no longer fit for marriage.

In fifteen Latin American countries, including Peru, rapists who agree (or in some cases merely offer) to marry their victims are exonerated of all criminal charges. Often a woman is pressured by *her family* to accept her assailant's offer as a way of restoring honor to both the victim and her relatives. In the words of a taxi driver interviewed by a reporter for the *New York Times*, "Marriage is the right and proper thing to do after a rape. A raped woman is a used item. No one wants her. At least with this law the woman will get a husband."

Not infrequently, rape occurs between relatives, especially when a father forces himself on his daughter. Although it is considered taboo by virtually all human societies, incest nonetheless occurs. Some have suggested that the widespread social prohibition of incest reflects its biological consequences: a pregnancy resulting from incest can produce genetically disadvantaged offspring. This phenomenon, known as inbreeding depression, burdens mostly the woman, who has a greater investment than does the man in any offspring that might be produced. But daughters can be easily overpowered by a determined father, whose own reproductive success could benefit from incest, and who—like a typical rapist—may force sex on his daughter when other reproductive avenues are blocked.

Mother–son incest is rarer. After all, young sons are presumably less able to overpower their adult mothers; by the time they are big enough to consider doing so, their mothers may be too old to be attractive to them. We believe that Freud was grievously wrong about the Oedipus complex. The desire and certainly the ability of sons to have sex with their mothers pale into insignificance compared with the desire and ability of fathers to rape their daughters.

Women do sometimes sexually molest men, but it is exceedingly rare. Anthropologist Bronisław Malinowski described a supposed exception in Melanesia, where a man, usually from a neighboring village, may be set upon and essentially gang raped by women.

> First they pull off and tear up his pubic leaf, the protection of his modesty and, to a native, the symbol of his manly dignity. Then by masturbatory practices and exhibitionism, they try to produce an erection in their victim and, when their maneuvers have brought about the desired result, one of them squats over him and inserts his penis into her vagina. After the first ejaculation he may be treated in the same manner by another woman.

Some anthropologists question whether Malinowski ever actually witnessed such an event, suggesting instead that his Melanesian informants were enjoying a good joke at his expense. We agree that the story is suspect. Not only would many men find being set upon in this way pleasurable, but the women also would be committing a biologically nonsensical act by mating with a man of uncertain quality who may be impregnating many women at the same time and who will then return to his own village.

Abusive Husbands

Domestic violence has spawned a network of safe houses, hot lines, support groups, and resource centers across the United States. Those seeking escape from its clutches are predominantly women; again, its perpetrators are overwhelmingly men. Moreover, they are men who are irritable, short-tempered, or jealous and possibly disinhibited by drugs, alcohol, or mental illness. Faced with an unpredictable barrage of emotional and physical abuse, many women find themselves in a terrible dilemma: leave and face economic dislocation or stay and endure abuse that is likely to continue. Many women, fearing that they will be more at risk if they leave the relationship, stay and try to manage a deteriorating situation. Yet when the physical abuse, substance abuse, or mental disorder worsens, as they frequently do, a violent outcome becomes progressively more likely.

It is not uncommon for a woman to stay with a violent man while laying plans for escape: sequestering money, arranging safe houses, consulting with lawyers on the side, documenting abuse episodes with visits to physicians and emergency rooms, and so forth. In her clinical practice, Judith has seen many women, who fear for their lives, in flight from their husbands or former lovers. One patient had no address because she was constantly on the move and had to make appointments via cellular phone.

Why do women choose such violent partners? To begin with, abusive behavior may not start until some months or years into a marriage. Before his dark side emerges, a man may display a controlling personality but one that also exudes power and caring. His aggression may appear in a positive light, as in a man who will passionately defend his wife, children, and property. Such assertiveness can be attractive, especially if the man is wealthy.

Some women think that "keeping the family together" for the sake of the children justifies staying in an abusive relationship, but rarely are children spared the devastating effects of a violent father, even if their mother thinks her submission buys stability for them. Yet women and children may well be better off economically living with a rich abuser than apart from him. It may also be that historically, women who mated with brutes gained some security against the ravages of other brutes. (One of anthropologist Napoleon Chagnon's many interesting findings about the Yanomamö is that women who were married to men with a reputation for fierceness were less likely to be abducted by other men.)

Furthermore, men's aggression toward their partners paradoxically sometimes *increases* their victims' emotional attachment to them. The Broadway musical *Oliver*, based on Charles Dickens's novel, *Oliver Twist*, offers a classic example. Nancy, a warm-hearted prostitute, is involved in an abusive relationship with Bill Sikes, brutal leader of a band of thieves. Nancy sings a beautiful love song, "As long as he needs me . . . ," after Sikes has viciously beaten her, suggesting that she wishes to be there for him because he needs (loves) her despite, not because of, his brutality. In real life, some women—typically those who suffer from low self-esteem—justify their husband's anger by telling themselves "If he didn't love me, he wouldn't care what I did," effectively convincing themselves that the man shows his love through abuse. Although such sentiments are perverse, the need to feel needed by one's lover frames many human relationships. From an evolutionary perspective, the man—terrible as his behavior is—has in fact made a decision to stay and on some level help rear his children. There are also instances in which women deliberately put up with or even provoke abuse from wealthy men in order to reap a sizable settlement in divorce court. Such cases are the exception, not the rule, but are an interesting example of how females may sometimes seek resources from males.

Regardless of why women stay with abusive men, it seems clear that brutish men appeal to some women, who may end up producing brutish sons, who in turn beget similar sons, and so on. In other words, sexual selection could easily have produced violence and brutality among men in the same way it has produced exotic and often bizarre traits among other male animals. Or perhaps, much as some women subconsciously choose to mate with sexy men as a way of producing sexy sons (as described in chapter 2), other women might mate with brutal men as a way of producing brutal sons, who would in turn experience reproductive success. In short, maybe we can add to the "sexy son hypothesis" a theory of "brutal bastards."

Feminist anthropologists Adrienne Zihlman and Nancy Tanner would disagree with such a notion. They argue that throughout history, women have preferred men who were decent and cooperative: "Females preferred to associate and have sex with males exhibiting friendly behavior, rather than those who were comparatively disruptive, a danger to themselves or offspring."

Maybe our ancestors did prefer kinder, gentler men, as many women do today. Maybe some were turned on by violent SOBs. But our guess is that—at least on occasion—brutes have forced themselves on unconsenting women and thus have kept their genes circulating from generation to generation.

Adultery

Just as male aggression and sexual behavior are closely linked, so is male aggression and sexual jealousy. The Old Testament speaks of a "jealous God," from which, even if we knew nothing else, we can assume it to be male, a god intolerant of sharing worship, of "having other gods," not just before this one, but alongside, after, or anywhere else at all. Husbands are particularly disturbed by the thought of their wives having other men, whether before them, after them, on the side, or any other way—yet at the same time, as we have seen, they may seek precisely such liaisons for themselves.

According to Kinsey and his colleagues, this double standard is both widespread and ancient: "Wives, at every social level, more often accept the non-marital activities of their husbands. Husbands are much less inclined to accept the non-marital activities of their wives. It has been so since the dawn of history." And, we add, it has been found to be so in every culture where such matters have been studied. Even in societies renowned as peaceable because they do not practice organized warfare—the Australian aborigines, !Kung bushmen, and most Eskimo groups—murders occur, and nearly all of them are sexually motivated, usually in retaliation for real or suspected adultery.

Historically and around the globe, adultery is viewed as a crime against *men*, that is, against the betrayed husband. So egregious is this crime that the male adulterer—the man who had sex with another man's wife—may be castrated or killed. In contrast, when it is the woman who is wronged—that is, when her husband commits adultery—there are few if any actual penalties, provided he is not "wronging" another married man in the process.

In ancient times, the Egyptians, Hebrews, Babylonians, Romans, Spartans, and others, defined adultery strictly by the marital status of the woman. If no man was "wronged," then essentially no wrong was supposed to have been done. When a married man slept with an unmarried woman, society generally winked, or looked the other way; by contrast, when a married woman slept with a man who was not her husband, the consequences were often catastrophic.

The theme recurs in literature and popular culture as well, from *The Scarlet Letter* to *The English Patient*. In Tolstoy's great novel of adultery, *Anna Karenina*, a liberal-minded gentleman named Pestov comments that the real inequity between husband and wife is that infidelity by each is punished differently (transgressions by the wife being taken more seriously). Anna's husband Karenin responds, "I think the foundations of this attitude are rooted in the very nature of things."

We had our own curious encounter with male–male sexual dishonor one year when we checked into a beautiful old hotel in Saas Fee, a picturesque town in the Swiss Alps. In the morning, David went hiking while Judith stayed behind to sleep late and read. The innkeeper knocked at the door under the pretext of fixing the toilet. Once admitted to the room, he lunged at Judith and attempted to rip off her blouse. Judith escaped his grasp and fled. When confronted later by David, the innkeeper apologized—to David—for attacking David's property (Judith) but not to Judith directly. From the innkeeper's perspective, it was David he had wronged; Judith was almost irrelevant.

Even during the French Revolution, when enthusiasm for creating a new society was so great that even the names for months of the year were tossed out and replaced with new ones, sexual asymmetries of the old regime were retained in one regard: legal sanctions against a *wife's* adultery remained in place as before. And in modern times, men are far more likely than women to cite adultery as a cause for divorce. In a sample of 104 societies, psychologist Laura Betzig found that a wife's infidelity was a primary cause of divorce in 48; a husband's infidelity, in not a single one.

Why does this double standard exist? The fact is that women get pregnant; men do not. An adulterous man may of course impose pregnancy on his lover, but so long as she is not married, no man is short-changed—that is, forced to rear children that are not biologically his. If an adulterous wife becomes pregnant, however, her husband may unknowingly help rear another man's offspring. (We return to the theme of "Mommy's babies, Daddy's maybes" in chapter 5.)

David once conducted an experiment on mountain bluebirds, small monogamous birds among which males and females cooperate to construct their nest and perform domestic duties, one working on the nest while the other forages. David observed mated pairs and, when the male was away, attached a model of a male mountain bluebird near the nest so that when the "husband" returned, he saw his "wife" consorting with a stranger. In each case, the male furiously attacked the model; not only that, he also attacked his mate, in one case driving her away.

When this experiment was repeated later in the breeding season, after the male's genes had been safely tucked inside his mate's eggs, the male responded with far less fervor. This simple experiment nicely demonstrates that aggressive intolerance toward suspected acts of adultery could very well contribute to a male's biological success.

The derivation of the word *adultery* is itself revealing: it comes from the Latin *adulterare*, meaning "to alter or change." To adulterate means to "debase by adding inferior materials or elements; making impure by admixture." The crucial admixture in this case is someone else's sperm.

Equally telling is the word cuckold. It comes from the European cuckoo, renowned for laying its eggs in the nests of birds of other species, which then become unwitting hosts. To add injury to insult, the newly hatched cuckoo chick ejects its host's biological offspring from the nest, thus monopolizing its foster parents' resources. To be cuckolded is to suffer the fate of those unwitting males who are oblivious to their wives' extramarital liaisons, ending up not only as biological failures but also as social laughingstocks. In *Love's Labour Lost*, Shakespeare gives us this cynical song:

> The cuckoo then in every tree;
> Mocks married men; for thus sings he,
> "Cuckoo; cuckoo"; O word of fear
> Unpleasing to a married ear!

When a man's parental investment is expended on behalf of another man's child without his knowledge, he may find that his love's labour is lost indeed.

Fear of women's infidelity, whether real or imagined, is sufficiently great among men to have given rise to innumerable strategies for guarding their mates. Even something as simple as an engagement ring given by a man to a woman sends a clear message to would-be suitors: this woman is taken. The more wives a man has, the more he must be concerned about their fidelity, as evidenced by the eunuchs whose function was to guard a sultan's harem or the court ladies of ancient China, who devised elaborate techniques to keep track of the menstrual cycles of the emperor's many wives and concubines.

Another Chinese technique was foot binding, which kept a wife housebound making it difficult for her to get away for extramarital sex or anything else. In other countries, especially India and Pakistan, women have been kept in purdah, or seclusion, isolated from men. In northern India, high-status women have been cloistered, virtually imprisoned in their own homes, as described by anthropologist Mildred Dickemann:

> You can tell the degree of a family's aristocracy by the height of the windows in the home. The higher the rank, the smaller and higher are the windows and the more secluded the women. An ordinary lady may walk in the garden and hear the birds sing and see the flowers. The higher grade lady may only look at them from her windows, and if she is a very great lady indeed, this even is forbidden to her, as the windows are high up near the ceiling, merely slits in the wall for the lighting and ventilation of the room.

Why do we consider such practices in a chapter on violence? Because in our view they represent one end of a complicated continuum in which males exert sexual control over and ownership of women, motivated ultimately by the economic and genetic consequences of infidelity. Thus, in many societies, wives are sequestered to guard against adultery and daughters are treated similarly to guard their virginity (which, in turn, ensures their marriageability and thus their economic as well as reproductive value).

One of the most notorious offshoots of this issue is the brutal and, by Western standards, grotesque practice of female circumcision, a potentially life threatening and often debilitating form of genital mutilation. Tens of millions of young girls, especially in East and West Africa, have been subjected to various forms of the procedure, which typically involves removal of the clitoris (clitoridectomy) and sometimes closure of the vagina (infibulation), which is accomplished either by sewing together the labia or by inducing severe scarring by cutting or burning the vaginal wall. An opening is left that is sufficient to allow menstruation but too small to admit a penis. When an infibulated woman marries, her vagina is cut open to permit sexual relations with her husband.

Although these practices are interwoven with much cultural tradition, their ultimate function is almost certainly biological, with men seeking to control the sexuality of women: clitoridectomy greatly reduces a woman's sexual pleasure and infibulation serves as a built-in premarital chastity belt. Women who refuse these procedures are liable to be socially scorned and marked as unmarriageable. Although it is nearly always older women who carry out the rituals, they are nonetheless acquiescing to the demands of men, to what they themselves were subjected to in childhood. Sadly, given the prevailing culture, it is in a mother's best reproductive interest to ensure that her daughters undergo these procedures; otherwise, they will be unlikely to have children themselves.

Murdering One's Spouse

"In every society for which we have been able to find a sample of spousal homicides," write Martin Daly and Margo Wilson in their landmark book *Homicide*, "the story is basically the same: Most cases arise out of the husband's jealous, proprietary, violent response to his wife's (real or imagined) infidelity or desertion."

Historically, in many cultures the murder of an adulterous wife or her lover has been not only condoned but encouraged. On the island of Yap (one of the Caroline Islands in the western Pacific), a cuckolded man "had the right to kill [his wife] and the adulterer or to burn them in the house." Among a tribe known as the Toba-Batak in Sumatra, "The injured husband had the right to kill the man caught in adultery as he would kill a pig in a rice-field." The Nuer people of East Africa recognize that "a man caught in adultery runs a risk of serious injury or even death at the hands of the woman's husband." These are not isolated, unusual cases; similar stories are found wherever human beings abide.

Such attitudes extend even to the state of Texas. Until 1974, homicide was fully legal there "when committed by the husband upon the person of anyone taken in the act of adultery with the wife, provided the killing takes place before the parties to the act have separated" (Texas Penal Code 1925, article 1220). It is not clear whether "in the act" meant that the husband had to slay his rival while sexual intercourse was literally in progress, but the idea is clear enough: a wife's adultery elicits such righteous wrath that her husband's act of murder is justified.

Only rarely do wives kill their philandering husbands. Significantly, when a wife does kill her husband, she most commonly does so in self-defense against *his* jealous rage, which arose because he discovered *her* affair. Although wives undoubtedly become upset when their husbands cheat on them, they are much less likely to respond with physical violence. Admittedly, Frankie killed Johnny because "he done her wrong," but in real life, men are more often the killers. Indeed, sexual jealousy is the second most frequent motive for homicide in the United States and Canada.

Even in the throes of the personal pain and anger caused by her spouse's sexual infidelity, a woman rarely pursues and kills a separated or estranged husband, yet it is distressingly common for a man to pursue and kill a separated or estranged wife. One man who stabbed his wife to death after they had been reunited following a six-month separation gave this account to the police:

> She said that since she came back in April she had fucked this other man about ten times. I told her how can you talk love and

marriage and you been fucking with this other man. I was really mad. I went to the kitchen and got the knife. I went back to our room and said were you serious when you told me that. She said yes. We fought on the bed, I was stabbing her. . . . I don't know why I killed the woman, I loved her.

Poet Carl Sandburg aptly summarized such conflict in a poem he dubbed a novel:

> Papa loved mama
> Mama loved men
> Mama's in the graveyard
> Papa's in the pen

There is some Darwinian logic in a man's responding murderously toward another man who has had sex with his wife; this is biological competition at its most intense. But why should the jealous husband react violently to his wife? In particular, why kill her? From a biological perspective, such behavior seems maladaptive in the extreme.

Three possible explanations present themselves. First, weak and ineffectual men are generally held in low regard. Like the Yanomamö men of today, who cultivate an image of fierceness, prehistoric men who responded murderously to their wives' infidelity may have been rewarded for doing so both socially and biologically. Second, if a man had multiple wives, his killing of one of them would send a powerful warning not only to the remaining wives but to other would-be adulterers as well. And third, there is the biologically bothersome possibility that a man will be unwittingly called on to raise someone else's child. Here is testimony in another tragic homicide case: "You see, we were always arguing about her extramarital affairs. That day . . . I came home from work and as soon as I entered the house I picked up my little daughter and held her in my arms. Then my wife turned around and said to me: 'You are so damned stupid that you don't even know she is someone else's child and not yours.' I was shocked! I became so mad, I took the rifle and shot her."

Add to these factors the despair often associated with low socioeconomic standing and the result can be a lethal brew. Most people, of course, don't kill their spouses or, indeed, anyone else. Murdered wives, for example, account for less than one ten-thousandth of 1 percent of the population. But when this population consists of several hundred million people, the result is several hundred murders per year, enough to give anyone pause.

Explanations Are Not Excuses

It is one thing to understand the male penchant for violence and entirely another thing to condone it. We consider violence to be the greatest problem confronting our species, whether it is directed toward women, men, children, larger social groups, or the natural environment. We do not believe that because violence comes more naturally to men than to women it is somehow pardonable, as if the fault lay entirely with evolution. It would be terribly perverse to use the material presented here to justify men's violence or to diminish its horror in any way. As Katherine Hepburn remarks sternly to Humphrey Bogart in the movie *The African Queen*, "Nature . . . is something we are put on earth to rise above."

At the same time, it serves no purpose to ignore the important sex differences that exist in aggressiveness and violence or to claim that these differences arise exclusively from faulty upbringing and ill-conceived social norms, although, to be sure, upbringing and experiences can exacerbate tendencies to which humans are already predisposed. When it comes to violent crime, social factors such as rage, retaliation, lack of positive role models, broken families, economic victimization, social despair, and outright ethnic and racial bigotry are undeniable. But evolution suggests that such factors only increase the likelihood that biological factors—*present in all people*—will be activated. No useful purpose is served by denying the existence of unpleasant things or sweeping them under the rug. (As linguist Deborah Tannen has suggested, this only makes for a lumpy rug.) Instead, let us employ biology's insights to make sense of why sperm makers are so prone to be troublemakers. On a more positive note, these findings may speak to the possible effect of enhanced socioeconomic opportunities for those—especially men—who are most excluded, disenfranchised, and alienated.

To some extent, it may be that violence, like poverty and disease, will always be with us. If such is the case, we have all the more reason to understand it.

CHAPTER 5

Parenting

My mother saith he is my father. Yet for myself I know it
not. For no man knoweth who hath begotten him.
— Telemachus, son of Odysseus, in
Homer's *The Odyssey*

Generations of human beings have been told, although usually not in so many words, that mothers
are the true nurturers, that fathers cannot be trusted
to rear children properly, and, moreover, that only by having
children can women fulfill their own basic needs. Such attitudes have spilled over into many a custody battle, with courts
of law even today far more likely to award custody to the
mother than to the father.

Indeed, an undeniable pattern exists the world over: women
in general and mothers in particular are the primary caretakers of children. Although men are certainly capable of child

care, not a single human society can be identified in which men are the primary parents. In fact, the index of a major scholarly volume published in 1991, *Cultural Approaches to Parenting*, has some forty-five page listings for "mother" and "maternal," but not one for "father" or "paternal." Nor is this bias unusual.

Everyone has heard the term *maternal instincts*, but how many have heard of paternal instincts? And although the term *working mother* has become a part of our everyday vocabulary, the term *working father* doesn't exist. Most of us simply assume that fathers work outside the home. Moreover, *working mother* refers specifically to women who work outside the home; caring for children full-time is generally not accorded the status of work, though it most assuredly *is* work, and difficult and demanding work at that.

If we are told that a woman mothered a child, we assume that she fed, clothed, cared for, hugged, loved, and consoled that youngster. But if we are told that a man fathered a child, we typically make no assumption beyond the fact that he inseminated the child's mother. A man can beget and forget; a woman typically cannot. Of course, it is undisputed that some fathers remain devoted even when forcibly kept at a distance and some mothers are indifferent to their children.

Although most fathers can change diapers, prepare bottles of milk or formula, and keep a watchful eye on their children as well as any mother can, study after study shows that fathers spend far less time with their children than mothers do. In the United States, employed married men interact directly with their children for an average of twelve minutes per day during workdays and for an average of twenty-seven minutes on their days off. Employed married women, by contrast, average about fifty minutes of direct interaction on workdays and, interestingly, less time (thirty-eight minutes) on weekends. Perhaps both the mothers and the children feel the need to bond after a long workday; it may also be that on weekends the fathers help out somewhat more.

Such patterns hold cross-culturally. For example, among the Ye'kwana people of the Venezuelan rain forest, fathers hold their infants on average only 1.4 percent of the time, whereas mothers do so 77.6 percent of the time (the balance is taken up by other relatives, nearly all of them female). In this disparity, the Ye'kwana are the rule, not the exception.

Even among societies in which fathers participate substantially in child rearing, gender differences are clear. The Aka pygmies of central Africa, for example, are unusual in the degree to which fathers are involved with their infants. Washington State University anthropologist Barry Hewlett, who lived among them, offers this observation:

> When the child wakes up at night and is not comforted by nursing, it is the father who sings to the infant and, if necessary, gets up and dances with the infant until s/he stops fussing. While fathers hold the infant, they are likely to clean mucus from the nose, pick lice from the hair and pick dirt off the body. If the infant defecates or urinates, he cleans up the mess. If the infant wants to nurse and the mother is not around, he offers his own breast to the infant. [He does not produce milk, however.]

Aka fathers hold their infants more, on average, than do fathers in any human society known to anthropologists; even so, during a twenty-four-hour period they do so, on balance, for only fifty-seven minutes. During this same time, mothers hold their infants for a whopping 490 minutes! Even in this society of "mothering men," women do eight and one half times more.

A survey of rural and nontechnological societies in such diverse places as Mexico, Java, Quechua, Nepal, and the Philippines underscores the skewed division of labor when it comes to parenting: fathers take care of their children 5 to 18 percent of the time (most commonly around 8 percent), whereas mothers pitch in 39 to 88 percent (typically 85 percent) of the time. The difference is great by any measure. Not only is it—as scientists are fond of emphasizing—statistically significant, but it is also socially, psychologically, and biologically significant. These male–female differences are even greater if passive forms of child care are included, such as watching one's children while doing something else or responding to them only when called. In such cases, mothers perform an average of about 15 minutes of child care for every 1 *minute* contributed by fathers.

Although single parenting by men appears to be on the rise, women nonetheless constitute about 90 percent of all single parents. It is also noteworthy that among those families in which men are the sole parent, only one in ten has children younger than five years of age.

Whether by preference or circumstance, the care of infants everywhere is overwhelmingly a woman's job. How is one to make sense of this aspect of sex?

Why the Imbalance?

Is it nature or culture that has created such an imbalance in the provision of child care? Most assuredly, men are every bit as capable as women of parenting in a modern society. If mothers and fathers have the same biological interest in their offspring, why do they not also have equal interest in rearing them? Why, on balance, are men less paternal than women are maternal?

One theory posits that differing roles for men and women go back to the early days of human history. Men, being larger and stronger than women, carved out their niche as hunters and leaders, leaving the less physically taxing chore of child care to women. Under such an arrangement, men would be socially, politically, and economically dominant over women, who in turn would be relatively powerless and oppressed. We suspect that there is some validity to this scenario; certainly women who devote themselves entirely to hearth and family are generally not the movers and shakers of the world. Full-time parenting unquestionably deprives women of powerful roles outside the home. Such powerlessness can be especially frustrating for women who find themselves unhappily married yet economically dependent on their husbands. The following true story is typical of this predicament.

Greg and Becky met at college, where both were good students. After graduating from college, they dated for several years while Greg went to business school and Becky earned her teaching certificate. After they married, Becky quit work to become a full-time mother to their two children. Although she had been trained as a teacher, she never appeared in front of a classroom.

In fact, Becky took on the role women had always filled in her family: she did all the domestic chores, chauffeured her children to their activities, and provided the support that enabled her husband to devote his full attention to work. During all those years Greg never cleared a table, ran a load of wash, or swept a floor. In short, Becky single-handedly met all the domestic needs of her family.

Partly as a result of Becky's devotion to the homefront, Greg rose rapidly in his profession, moving into management and then into upper management. He worked as hard at business as Becky worked at home. As her husband became more successful and more powerful, Becky felt increasingly isolated and became anxious and depressed. She found solace in food; in fact, she gained 120 pounds.

After twenty-seven years of marriage and numerous extramarital affairs, which Becky half-consciously acknowledged, Greg announced that he wanted to divorce Becky in order to marry a vivacious career woman twenty years younger than himself. In the divorce negotiations, Greg resisted paying alimony, pointing out that he and Becky had the same level of education. Becky's attorney was quick to counter that she hadn't worked outside the home in twenty-six years, her credentials were outdated, and her earning capacity was a small fraction of Greg's. The emotional issue—Becky's dependence on Greg—was almost an afterthought. Abandoned by her husband, and with her children grown and gone, Becky's personal identity—that of homemaker—had disintegrated.

As Becky's case reveals, parenting can generate a vicious circle, with each step reinforcing the other: the more women do the child rearing, the more socially and economically powerless they become. And with generally less education, fewer career options, and less money, women eventually become mired in their powerlessness: their lack of options makes them less able to do anything but child rearing, which makes them yet more dependent on their husbands.

Radical feminist Shulamith Firestone may have been right when she observed that "the heart of woman's oppression is her childbearing and childrearing roles." Where Firestone erred, in our opinion, was in her assumption—widely shared—that these roles are determined by social forces alone and that women are simply strong-armed into childbearing and child rearing by churlish men.

The Working Woman's Dilemma

Many women, of course, aspire to being more than just mothers, so they divide their time and energy between children and careers. But as nearly every one of these women can attest, it is devilishly difficult to care for children and at the same time pursue a successful career. Even when husbands share the chores of homemaking and parenting, working mothers often feel terrible frustration, continual fatigue, and a gnawing sense of dissatisfaction.

Judith sees many professional women in her practice, and the story is nearly always the same. They are bright, energetic, competent women who want it all: career, money, and children. But they seek therapy because they are exhausted, often depressed, and confused. Even with full-time nannies and housekeepers, they feel unsupported. Even with devoted husbands who are good fathers, they are angry. It is simply painful and frustrating to be a mother at a distance, knowing that someone else is going to the school plays, the teacher conferences, the orthodontist appointments. Something inside whispers that good mothering means cuddling and baking cookies, not being the breadwinner.

A study by Norma Radin, a sociologist at the University of Michigan, supports the assertion that at some level many women *want* to be the primary caretakers of their children. Radin examined middle-class families in Michigan, all of whom were committed to egalitarian child rearing, that is, child care in which the fathers played a substantial role. Among these couples, mothers complained that they didn't have enough closeness and involvement with their children, whereas fathers groused about being hampered in their careers. It is possible that these findings reflect society's different expectations for men and women. For her part, Radin concludes that "even when parents choose to violate sex role expectations, there are still internal pressures to fulfill the tasks for which they were socialized." We suggest, however, that prior socialization is not the entire answer and that the "internal pressures" are internal indeed.

Judith spent several years helping a patient work through her turmoil over mothering. When Jessica, a systems analyst, first met Roger, a professor, she was delighted by Roger's parenting style. Roger had sole custody of two young daughters from a previous marriage. It was not uncommon for Jessica to enter Roger's home and find him at the typewriter writing a grant while his children watched *Sesame Street* on television and dinner bubbled on the stove. Jessica was attracted by Roger's ability to diaper a two-year-old and write part of a research article in the same ten minutes. Roger was very comfortable with his style of parenting, which he referred to as benign neglect: he attended to his children's basic needs but otherwise focused on his academic pursuits.

Jessica and Roger eventually married and had a child together. Ironically, the very behaviors that pleased Jessica while they were dating in-furiated her once they had a baby. Roger was quite content to stay at home, baby in the Snugli, writing articles. He cooked every night, made sure his older children got to and from school, and occasionally ran a load of laundry. But Jessica soon became critical of Roger's lais-sez-faire style and chastised him for his work habits. What especially bothered her was not Roger's behavior but the fact that her mothering role had been usurped. When Roger walked into the room, the baby's eyes would light up; when the baby bumped herself, it was to Roger's comforting arms she went; not Jessica's. Jessica felt unneeded and didn't like it.

Although most career women who become mothers do not face the same issues as did Jessica, they are troubled in other ways. Many are simply exhausted at the end of the day, when the demands of part-time parenting and a full-time job add up to an impossible schedule. Says one friend of ours, "I'd love it if my husband would put the kids to bed, but they won't have any part of it. I can't say that I blame them . . . Jack is so task oriented: get the kids in bed, read a story, turn out the light, and that's it. They want me because I linger, cuddling and giggling with them until they fall asleep." But she's quick to acknowledge that the bedtime ritual exhausts her, and she knows life would be easier if she didn't also have to get up at dawn and go to the office.

Barbara epitomizes the angst of a young working mother. She wants to nurse her daughter, Abby, until she is two, but she also hopes to climb the corporate ladder at the law firm where she works. Barbara rises at 5:30 a.m. to nurse Abby and then hands Abby over to her nanny so she can be at work by 7:00. Every three hours, Barbara secludes her-self in the rest room to pump her breasts, a task that keeps her away from her desk for fifteen minutes at a time. At 5:00 p.m. Barbara leaves the office, arriving home around 6:00 with full breasts as well as bot-tles of milk for the next day. She then spends the evening feeding, play-ing with, and bathing her daughter. Barbara notes that whereas Abby seems glad to see her in the evening, she is equally happy to see the nanny each morning. Although Barbara is thankful that Abby is well cared for, she envies the nanny and is considering working part-time, though doing so would scuttle her chances for promotion.

Few men agonize over such matters; in fact, many rejoice in the op-portunity to go to work and escape the daily grind of child care.

Back to Biology

Androgynous parenting is a long-standing dream, yet large-scale social experiments attest to the depth—and biological stubbornness—of sex differences in parenting behavior. In the 1950s, for example, Israel launched the *kibbutz* movement. It was a courageous attempt to establish an egalitarian, nonsexist society: men were to partake equally in the traditionally female roles of child rearing and domestic responsibilities, and women were to be equal participants in organizational and other typically high-ranking male tasks. For the first few years, the movement seemed a roaring success.

Twenty-five years later, however, the revolution was over, its goals upended. Kibbutz women had reverted to domesticity, and men were running the affairs of the group. Contrary to the movement's hoped-for egalitarianism, boys played aggressively, imitating heroes and fierce animals, while girls doted on dolls and pretended to be mothers. Women, wanting more time with their children, opted to do the bulk of the parenting. In his book *Gender and Culture: Kibbutz Women Revisited*, anthropologist Melford Spiro of the University of California at San Diego describes this outcome as "the triumph of nature over culture."

Further evidence for the persistent biological "nature" of parenting comes from a well-respected study of six widely separated rural and traditional cultures conducted by Beatrice Whiting and John Whiting of Harvard University. In every one of these cultures, Whiting and Whiting found that girls were far more likely than boys to spend time with infants and to behave nurturantly toward them, a difference that increased from ages three to eleven. Of course, just about every culture on record tends to push girls toward nurturing interactions with young children and to encourage boys to be if not antagonistic toward those younger than themselves, at least more distant. Nearly everywhere, girls are expected to care for and be interested in infants and younger children. Baby-sitting, caring for siblings, playing with dolls, and the like are very much girls' activities. To be sure, social pressures contribute to these observed sex differences. But the universality of these pressures suggests that they are the result, not the cause, of male–female differences.

Interestingly, studies also show that seasoned fathers are no more involved with their second child than they were with their first. In fact, they spend relatively more time with the eldest child, who is likely to be walking, talking, more responsive to play, and less in need of intense caretaking. Differences in experience, in short, do not seem to explain male–female differences in parenting.

Other evidence comes from studies of human emotions. Researchers have found that women are more likely than men to show empathy, affiliation, social skills, and sensitivity to nonverbal cues as well as greater attention to cuddling and meeting a child's immediate needs. Women are also better at detecting emotions, as borne out by a study of the ability of university students (twenty men and twenty women) to recognize the facial expressions of infants. When close-up photographs of the faces of babies were flashed on a screen, the subjects were asked whether each showed joy, surprise, interest, sadness or distress, anger, fear, or disgust. The results? Women were significantly more accurate than men in their ability to identify emotions; they were also faster in their responses. It is noteworthy that previous experience with infants and children had no effect. Apparently, it isn't just being a parent that brings out a person's ability to "read" an infant but also being male or female. These findings are consistent with other research indicating that mothers are better than fathers at recognizing the cries of their children and also at interpreting the meaning of those cries (pain, hunger, and so forth).

Women's sensitivity to the nuances of facial expression, whether of infants or adults, may be an offshoot of female choice, which requires that females have a discriminating eye when it comes to mate selection. Women who are especially acute in social judgment and assessment, with an ability to "see through" words and assess nonverbal cues, would be more apt to distinguish a good potential mate from a bad one. Or such sensitivity may result, in part, from the well-known phenomenon in which social subordinates are especially adept at interpreting the mood and intent of their superiors. If so, either scenario would make female sensitivity to nonverbal cues an example of what Harvard University paleontologist Stephen Jay Gould calls a "spandrel," a nonadaptive by-product of selection for something else.

Most likely, however, natural selection would have directly favored women who showed sharp judgment when it came to child rearing. In a prehistoric setting, a mother's sensitivity to her children—especially when it came to their meeting basic needs and caring for them when they were sick—would surely increase their likelihood of survival, which in turn would lead to descendants who were similarly attuned and responsive. All a father had to do was defend his family and ensure that they had enough to eat. Tenderness could be left to Mom.

In fact, philosopher and pacifist Sara Ruddick proposes in her recent book, *Maternal Thinking*, that the hands-on experience of child rearing gives mothers an abiding sensitivity for the vulnerable, developing human body, a tenderness that somehow eludes men. Because of that exposure, she says, women are naturally compassionate and thus are more likely than men to promote pacifism and a "caring" economic and social policy. If true, this might in itself be a reason for men to do more parenting.

When Men Parent

Just as they bring their maleness to lovemaking or their competitiveness to the tennis court, men bring their own style to parenting. Most women are "softer" than men in their interactions with their children, being more likely to hold and comfort them, listen to their stories, and wipe away their tears. In contrast, men tend to be "harder" in the sense of being more physical, that is, more boisterous, more inclined to carry their children piggyback or toss them around playfully than to hold and kiss them. They also tend to be less patient with their offspring.

Toddlers, for their part, often differentiate between their parents. They'll go to their fathers to play but seek their mothers to get their basic needs met. Daddy is a treat; he means fun. Mom is more gentle but, when it comes to the basics, more serious.

Alice Rossi, a highly respected sociologist and former president of the American Sociological Association, believes that biology plays a significant role in the parenting inclinations of men and women. She describes the situation of a man she calls Stuart, a history professor who cared for his newborn son four mornings a week while his young daughter attended nursery school. According to Rossi, things went well for the first few months because the baby napped most of the morning and Stuart could spend two or three hours preparing lectures. Once the baby began sleeping less, however, Stuart reported that he had trouble comforting him. When Rossi asked Stuart about his feelings under such circumstances, he said that sometimes "I go pound my fist on the wall or something like that."

At the same time, Stuart reported feeling good about his young daughter, telling Rossi, "My older child now is verbal . . . she dresses herself, takes care of herself, goes to the bathroom by herself, everything, a more or less autonomous being . . . and I just enjoy that tremendously." Obviously, the daughter's skills in taking care of herself reduced the need for caregiving by the father and allowed him to get on with his own work.

For Stuart fathering meant being around his children but free to pursue his own interests. Asked what he did when the baby was awake, Stuart said, "I try to do something constructive still, maybe a little reading or some project around the house . . . sometimes I'll be in the same room with him; other times I'll just let him play by himself."

Stuart's parenting style could be described by the sociological term *role distancing*, as opposed to *role embracing*. Like Roger, who practiced benign neglect, Stuart was exceptional in being *unusually involved* with his infant; he and his wife had decided to make child rearing as egalitarian as possible. And yet in the actual care of his infant, Stuart—like most men—was a distancer.

Also interesting is that men commonly feel the need to distance themselves physically from their children, often pressuring their wives to hire a baby-sitter so the couple can go out. For a man, the opportunity to leave young children at home and go to a restaurant or a movie represents freedom; for a woman, it is more likely to elicit conflicting emotions. She, too, wants to go out and have a good time, but she worries that the baby may be lonely, frightened, or even mistreated and on some level may believe that she is not a good mother to abandon her child in this way. Most men do not suffer from such ambivalence. In fact, a husband may add to his wife's emotional turmoil by reacting negatively to her anxiety, interpreting it as a sign of her lessened commitment to him.

Mommy's Babies, Daddy's Maybes

Sex roles in parenting need not be fixed. For proof, just take a look at the bewildering array of parenting arrangements throughout the animal world. In some species, females care for their young; in others, males do. In yet others, males and females parent together. There are even cases, such as the European cuckoo, described in chapter 4, in which parenting duties are left to an entirely different species. And many animals exhibit no parental care at all. Most insects, for example, simply abandon their eggs after they are laid, leaving them to fail or flourish on their own. But the norm—at least among birds and mammals—is for one or both parents to care for their young until they can make it by themselves.

Indeed, the intensity with which a parent guards and cares for his or her young is surprisingly great among some frogs, reptiles, and even insects. Male gladiator frogs, for example, dig the nests in which fertilized eggs are laid. Aptly named, they vigorously defend their eggs until they hatch, wielding needlelike spines at the base of the thumb to slash the skin and eardrums of any intruder.

Parental care reaches its peak intensity, however, among mammals. Females typically nurse their young for weeks or months after birth. A fawn may suckle its mother for about four months; a whale pup, for two to thirteen years. Among humans, mothers frequently nurse their babies for a year or more; some will even do so intermittently for three or four years. Equipped as they are with a built-in food supply, it seems only logical that female mammals would do the bulk of the parenting.

Yet, bizarre as it might sound, there is no biological reason why male mammals cannot breast-feed their young. They have the necessary equipment: nipples and rudimentary breasts. They even have the physiological potential to produce milk. But they don't do so because the hormones that stimulate breast development and milk production are suppressed by the male brain. When those hormones are released, as sometimes happens as a side effect of certain medicines, some men do in fact develop breasts and secrete milk. There is no a priori reason why evolution couldn't have designed a parenting plan in which males lactate, stimulated perhaps by witnessing a birth or simply by smelling or touching a newborn baby or hearing its cry. Natural selection has, after all, created more improbable scenarios.

But lactating males don't exist (except for an unconfirmed report in one species of Malaysian fruit bat). At stake seems to be the high cost of milk production weighed against confidence or, rather, *lack* of confidence, of paternity. Simply put, children are inescapably "Mommy's babies, Daddy's maybes." Whereas every mother naturally bears an ironclad relationship to her child, no father can ever be entirely confident—short of keeping his wife in solitary confinement, shackling her with a chastity belt, or demanding DNA testing—that he is the real (that is, the genetic) father. Here we have one of the great asymmetries of the natural world. Recall the uncertainty of Telemachus, expressed in this chapter's epigraph. "Telemachus's lament" also works the other way: no man knoweth for certain whom he hath begotten.

In his play *The Father*, Swedish writer August Strindberg describes the dilemma of a husband tormented by uncertainty as to whether he is the biological parent of his child: "I know of nothing so ludicrous as to see a father talking about his children. 'My wife's children,' he should say. Did you never feel the falseness of your position, had you never had any pinpricks of doubt?"

Such pinpricks are occasionally called for. An obstetrician friend of ours, for example, found himself in the awkward position of presenting a couple—two blond, blue-eyed Caucasian parents with their new curly-haired, dusky-skinned baby. The husband, eagerly anticipating the birth of his first child, had had no inkling that his wife had been unfaithful and perhaps would never have known of her transgression if the evidence—a biracial baby—had not been so notable.

Friedrich Engels, one of the fathers of socialism, believed that early men went to great pains to ensure their paternity. In *The Origin of the Family, Private Property, and the State*, Engels proposed that this universal problem of uncertain paternity gave rise to the very notion of the human family itself. "The family," wrote Engels, "is based upon the supremacy of the man, the express purpose being to produce children of undisputed paternity." But his view was based on economics, not biology. "Such paternity," he wrote, "is demanded because these children are later to come into the father's property as his natural heirs." A biological perspective would reverse Engels's vision of cause and effect: the reason property is preferentially given to one's "natural heirs" is that these individuals are genetic relatives and thus are biologically important to the giver, in precisely the same sense that milk is given to a child who is a genetic relative and thus is biologically important to the donor . . . in this case, the mother.

Guaranteeing Paternity

The cost of raising another male's young is so steep that ingenious ploys to guarantee paternity crop up throughout the animal kingdom, especially among species in which fathers care for their young. Males of the giant water bug species *Belostoma flumineum*, for example, incubate the fertilized eggs, carrying them faithfully on their backs and aerating them until they hatch. During this time, the males don't eat; instead, they devote themselves to caring for "their" eggs. A male Belostoma is unlikely to be duped into carrying eggs fertilized by another male because he copulates with the female immediately before she deposits those eggs on his back. In fact, he repeatedly interrupts the egg laying to mate yet again. In one observed case, the male interrupted his mate's egg laying more than 100 times over a period of thirty-six hours. By exerting such tight control over the process, the male water bug takes no chances that he will be caring for someone else's offspring.

Among some species of walkingstick (*Diapheromera*) the male, which is much smaller than the female, rides on the female's back, his genitalia linked to hers, until the eggs are laid. In this way he functions as a living chastity belt, closing off the female's reproductive tract with his own genitalia so no other male has access to it. The couple may remain linked for several weeks.

Other strategies for ensuring paternity among invertebrates (some of which were discussed in chapter 2) include cementing closed a female's genital opening as a way of blocking other males' sperm from reaching her eggs, as seen in certain species of parasitic worms; turning one's penis into a scoop in order to remove a predecessor's sperm before depositing one's own, as occurs in at least one species of damselfly; and riding piggyback on the female as a way of warding off would-be suitors, as has been observed in many beetle species, including the iridescent Japanese beetles that devour North American gardens. A male honeybee makes the ultimate sacrifice: his body explodes like a grenade, which helps entrench his sperm within his mate's genital tract.

In the great majority of birds, males provide parental care. It seems that in such cases, uncertain paternity is overridden by ecological factors: the helpless hatchlings grow so quickly that a mother acting alone simply cannot provide enough food. So the father helps too, apparently presuming that the hatchlings are his. If he did not, his evolutionary success would be about zero.

Among mammals, polygynous males are less paternal than monogamous ones. There are probably two explanations for this dichotomy. First, having multiple mates cannot help but limit the amount of parenting any one male can undertake; after all, how could an elephant seal with some forty pups give any of them much attention? Second—and more important—a polygynous male, assuming he has no eunuchs to serve him, cannot control his mates as well as a monogamous male can. Put another way, a harem member probably can get away with a little sex on the side more readily than can a female in a monogamous relationship. Thus, paternal behavior on the part of a harem master may be time wasted if he isn't certain that the offspring in his entourage are his.

In one telling experiment, harem-keeping red-winged blackbirds were vasectomized and thus rendered sterile. Yet "their" females continued to lay fertile eggs, which developed into normal offspring. Apparently, the females were having sexual liaisons with nearby bachelor males. On the basis of this information, it may not be surprising that the average male blackbird rarely feeds "his" children, perhaps because somewhere deep inside his feathery head he senses they aren't necessarily his.

In almost all species, the degree of parenting by the male reflects the confidence of his genetic ties to the young. This is not an intentional, self-aware confidence—although to be sure, such assessments have occupied the conscious minds of many generations of human beings—but rather the dark, dim biological stirrings shared by much of the animal world. Animals have no more need to understand the evolutionary genetics of relatedness than a rock needs to understand the law of gravity in order to "behave" in accordance with it.

Significantly, paternal behavior among mammals is almost always associated with strict monogamy, in which males can be reasonably assured that they—and not some interlopers—are the fathers of their offspring. Male beavers, foxes, and coyotes are all pretty good providers and caretakers of their young, and it is no coincidence that they are all monogamous. Among these species, confidence of paternity is high and males lavish attention on their young.

For evidence, consider the marmoset, a monogamous primate that demonstrates an unusual degree of paternal investment. The male pygmy marmoset, for example, actually assists in the birth of his offspring, holding each newborn until the next one emerges from the mother. (Marmosets typically bear twins.) During the first week of his infants' lives the father chews their food for them. He also carries them during the day, bringing them to their mother every few hours for nursing and then retrieving them afterwards, until they are about three months old. Even after the young are weaned, monogamous male marmosets continue to carry them about.

In contrast to the marmoset, another primate, the male East African baboon, shows few paternal tendencies. Fathers tolerate their infants but do almost nothing for them. However, these animals live in multimale troops, and so, unlike the case with marmosets, relatedness between any given adult male baboon and any given infant is less than certain. Still, research shows that if a male has consorted sexually with a particular adult female several months before the birth of her young and thus has a chance of being their father, he is more likely to behave solicitously toward them.

Rutgers University ornithologist Harry Power showed what can happen when a male animal that is definitely *not* the father associates with a female that is raising offspring. Experimenting with mountain bluebirds, Power set out nest boxes and waited for pairs of bluebirds to move in, mate, and produce young. Shortly afterward, he removed the males, which normally would have helped rear the nestlings, bring them food, chase away intruders, and perform other paternal duties. In their place came new male bluebirds, which readily moved into the nest boxes, almost certainly attracted by the prospect of mating with a wealthy widow. Stuck with offspring not their own, these stepfathers did virtually nothing to assist in rearing the young bluebirds. Only one in twenty-five helped feed the nestlings, and none gave alarm calls in response to predators.

A variation on this theme has been observed among dunnocks, rather ordinary looking European songbirds whose sexual and reproductive behavior belies their drab appearance. In that species, a socially dominant male jealously guards his mate during the breeding season. Although dunnocks, like most birds, are usually monogamous, occasionally the female gives her "alpha" male the slip and mates with a subordinate, "beta" male. When the youngsters hatch, they are inevitably cared for by their mother, assisted by the alpha male. Sometimes the beta male also helps out, but significantly, he only does so if he had previously copulated with the female. In fact, when two males are present, they adjust their caretaking to correspond with their prior sexual access to the mother. This finding is doubly important: not only does it reveal how behaving paternally is keyed to confidence of being genetically paternal, it also suggests a possible reason for the female dunnock's sexual interest in more than one male—namely, to obtain parental assistance from them both.

There is yet another intriguing aspect of the dunnock's sex life. If a male discovers that his female has been spending time near another male, he pecks vigorously at her cloaca. She responds to her current consort's peck by obligingly squeezing out a few packets of his predecessor's sperm. And this, in turn, increases the likelihood that he will father at least some of the eggs to be laid.

The overall pattern is clear: genetic relatedness strongly predisposes an individual to provide parental care. Further evidence comes from species such as fish that practice external fertilization. In the great majority of such cases, male and female both release gametes into the sea and fertilization takes place in the surrounding water. As a result, neither male nor female can be absolutely confident of being a genetic parent. And so, just as evolutionary theory would predict, male fish are about as likely as females to act like a parent. This may involve aerating the eggs, keeping them free of fungus, defending them from predators, and sometimes feeding the young fry.

When, however, fertilization takes place inside the female's body, female confidence rises but male confidence plummets, simply because the male can never be certain he is the only one that has inseminated his mate. As would be expected, along with a decline in confidence of paternity occurs a parallel decline in the male's participation in care of his offspring.

Here, then, is a key point for human beings: internal fertilization sets the stage for mothers to be more maternal and fathers to be less paternal. In addition, humans carry unmistakable signs of their polygynous heritage, a fact that further lowers a male's confidence in the sexual fidelity of his mates. Add these together and the stage is set for dramatic male–female differences when it comes to parenting.

Reassuring Dad

Among humans, the issue of paternity manifests itself right from the start—at the birth of a baby. Psychologists Martin Daly and Margo Wilson of Canada's McMaster University recorded the spontaneous statements of parents after 111 births in the United States and Canada and then polled the parents of another 526 Canadian infants. They found, first, that paternal resemblance elicited many more comments than did maternal resemblance: "He has his father's chin" or "she has her father's nose." Rarely was it said of an infant, "He has his mother's ears" or "She has her mother's mouth." This focus on paternal resemblance is a familiar and widespread phenomenon. When, in the musical *Oklahoma!*, Ado Annie—"the girl who can't say no"—asks her fiancé about having a child, he replies, not missing a beat, "He'd better look a lot like me!" Such comments are just what would be expected when maternal resemblance—or, rather, maternal relatedness—is taken for granted, whereas the situation for fathers is more questionable.

Daly and Wilson further noted that it is primarily mothers and their relatives who make these claims of resemblance. There should be no surprise here, either: by boosting the husband's confidence in his paternity, the mother's side heads off the possibility of abandonment. Even so, fathers tend to be somewhat skeptical about such claims, whether from modesty or from genuine uncertainty. Similar results were obtained in a study of Mexican infants and their families, a finding that suggests the concern over fatherhood is cross-cultural.

The Art of Deceit

Men are doubtless capable of being good fathers, if that is in their interest. They—like other primates—are also capable of being rather indifferent fathers. Furthermore, they can be quite deceitful about their intentions. Primatologist Barbara Smuts, for example, found that among olive baboons, adult males and females commonly form friendships that involve sleeping together, grooming each other, and sharing food. When the male interacts benevolently with his female friend's offspring (a relatively rare event), it seems he may have other motives: male olive baboons that behave solicitously toward a female's offspring significantly increase their chances of mating with that female.

Vervet monkeys are even more devious. In one experiment, when an infant's mother could be seen, males behaved nicely toward the youngster, but when the mother was not visible, their tolerance plummeted. Mothers in turn reacted positively toward those they had seen be nice to their infants and behaved aggressively toward those they had seen be mean.

In his book *The Selfish Gene*, British zoologist Richard Dawkins contrasted "dads" and "cads." Dads put their emphasis on the promise of paternal care and investment, whereas cads try to achieve matings, with essentially no behavioral follow-through. It might seem that females would always prefer dads, since their offspring would then have better care. But part of the strategy of successful cads seems to involve deceiving their would-be mates as to their actual inclinations. In short, sexy, dashing cads may succeed by pretending to be dads.

One anthropologist, observing social dynamics in a rural village in Trinidad, found that when a single mother had a child, a prospective husband interacted more with the child before marrying the mother than afterward. Not unlike male vervet monkeys, Trinidadian men cozy up to women by being nice to their children, giving the impression that they will be better stepfathers than they turn out to be.

Stepparenting

Anyone who has inherited stepparents or stepchildren or been a member of a blended or broken family has experienced firsthand the complexity of nongenetic parenting. When single parents date each other great effort is often expended to include the children. The dating phase may be a honeymoon for the entire potential stepfamily, with both adults going out of their way to court their lover's children. It is not uncommon, for example, for the families to go on joint vacations to Disneyland, plan fun-filled trips to the park, or splurge on new toys. The more delighted the children are with the potential stepparent, the greater the chances of a wedding.

Once the marriage has taken place, however, the heady tolerance of courtship can quickly dissipate. Sentences such as "He's your son, not mine, so you take care of it" or "You're not my mother; you can't tell me what to do" are commonly uttered. Fueling the tension, of course, are the conflicting emotions of the two parents, who want to be fair and to maintain their marriage but who also desperately want to protect their biological children from insensitive treatment by stepsiblings and, often, by a less-than-devoted stepparent.

We have experienced these struggles firsthand. When we first met, Judith's five-year-old son entertained David's offspring with his various toys. The children were happy—indeed, fascinated—with one another.

We seemed to have the perfect "new" family. Months of trips to the community swimming pool, hiking excursions, and eventually a sojourn to Maui followed. Our first two years of courtship and marriage were halcyon times.

Our bliss abruptly ended, however, with the birth of our first "shared" daughter. Suddenly, Judith found David's offspring loud and demanding and David found Judith's messy and insolent. Fights developed over private versus public schools, religion, dinner rituals, and chores. Sadly, we cannot claim to have found a happy solution. To this day, twenty-two years later, stepfamily problems top our list of failures and frustrations. Inequalities in child support, inheritance, family expectations, and personal styles have made life truly miserable . . . at times.

In some families, clear lines are drawn, with each parent agreeing to be the primary disciplinarian and caretaker of his or her own children; in other families, a decision is made to treat all children equally in hope of having a harmonious household, although such harmony is rarely achieved. In many ways, the fairy tale "Cinderella," though grossly exaggerated, embodies the conflict created by blended families. In this apocryphal tale of a mistreated stepchild, the abusive parent was a stepmother intent on appropriating Cinderella's father's wealth for her own daughters. Sadly, both men and women are adept at deploying deceptive strategies to gain advantage for their own offspring at the expense of a stepchild.

When (Step) Parents Kill

One of the most horrifying perversions of parenting is the killing of infants, an act known as infanticide. For many years, infanticide was thought to be unique to human beings, a pathological human behavior reflecting unbridled rage toward a misbehaving child. In the 1970s, when biologists first documented infanticide among animals, the scientific community was skeptical. The researchers were mistaken, critics said. Even as proof mounted, skeptics claimed that infanticide must be an aberration, perhaps a result of overpopulation or some other pathological condition. Over the past twenty years, however, the painful truth has slowly dawned: as part of the normal lives of many animals, adults kill infants. And most of the time the killers are adult males. The cold fact is that when males kill infants, especially those not their own, they often further their own genetic success.

Overwhelmingly, when men and women kill children (as when other animals do), the victims are most likely to be stepchildren or strangers with whom the killer has no genetic connection. So strong is the correlation that homicide detectives immediately look for the presence of a nonrelative in a murdered child's life—especially if that person is sexually involved with the child's mother.

In the large number of animal species in which infanticide has been observed, the scenario is nearly always the same: one male usurps another and then kills the offspring of his predecessor. Rodents do it; even the ostensibly noble lions do it. Primatologist Dian Fossey (immortalized in the film *Gorillas in the Mist*) described infanticide among usually peaceful gorillas, with a bereaved female eventually mating with the same male that killed her offspring.

Some of the best-documented examples of infanticide among animals come from the langurs of India: lovely, slender, silver-gray monkeys that typically live in harems consisting of a dominant male, several adult females, and their offspring. As in most polygynous species, this arrangement excludes a number of males, many of them subadults not yet big enough to displace the alpha, or dominant, male. These resentful bachelors congregate in loosely organized bands, periodically attempting to kick out the alpha male and take over his females. Eventually, when the harem master is overthrown, the newly ascendant bachelors fight among themselves, after which one of them settles down as sole proprietor of the langur females and their young.

With meticulous cunning and stubborn persistence, the new alpha male then proceeds to stalk and kill the young monkeys in the group; the process can be grisly and gruesome, with infants dying slowly from numerous bites. Anthropologist Sarah Hrdy, who has studied infanticide among free-living animals more than has any other scientist, once divulged that after witnessing an especially gruesome killing (in which the victim and its mother were well known to her), she wept.

Sympathy and sentimentality aside, infanticide represents a smart evolutionary strategy for the male langur. Once their babies are killed, the mothers in his harem cease lactating and begin ovulating, whereupon they mate with their infants' murderer, producing a new round of offspring, which, not surprisingly, he tolerates. By his actions—dastardly as they are by human ethical standards—the male langur perpetuates his genes at the genetic expense of his predecessor (and that of his mate). At the same time, by her actions—cowardly and callous as they seem—the female langur is doing the best she can to increase her own biological success. Once her infant is dead, the langur mother is faced with a stark choice: to breed or not to breed, to maximize her fitness or not. By mating with her child's killer, she has also chosen a dominant male, which presumably has good genes and the ability to fend off other male intruders, at least for a time.

In some cases, then, natural selection has clearly favored infanticide. Although its evolutionary origins seem clear, little is known about what produces the actual impulse to kill a helpless infant. Preliminary studies suggest that hormones are involved. For example, if normally monogamous male birds are given extra doses of testosterone, they suddenly display unusual aggressiveness toward other males, and as a result they accumulate more than one female on their breeding territories. Because these males father more offspring than do normal, monogamous males, the question then arises why natural selection did not favor higher levels of testosterone to begin with. The answer is that testosterone-enriched males show a dramatic reduction in parental care, and as a result they have lower reproductive success than normal, monogamous males that assist their mates in rearing offspring.

Excessive levels of testosterone are also known to influence parental behavior among mammals, nearly always for the worse. Male mice, for example, are more likely than female mice to kill strange pups, but when females are given testosterone, they become killers as well. A female rabbit dosed with testosterone during the twelve days before she gives birth is more likely to scatter her pups after they are born, less likely to nurse them, less likely to build an adequate nest for them, and more likely to kill and eat them. In one closely observed troop of semi-wild rhesus monkeys, only a single adult male behaved in a genuinely nurturant way toward infants, and this was the one male that had been castrated.

Mothering behavior, in turn, is known to be affected by the female hormones prolactin, progesterone, and oxytocin. Recently, the presence of a particular gene known as fosB has also been linked to maternal behavior. Led by Michael Greenberg at Harvard University's School of Medicine, researchers produced a strain of mice lacking this gene. Although these mutant mice seemed normal in every other way, the females were effectively infanticidal: they abandoned their newborns at birth, thus dooming them.

Summing Up

Nothing we have said, read, theorized, or experienced suggests that fathers are incapable of loving, and loving deeply; nor have we any reason to deny that sometimes mothers are terrible parents, occasionally even homicidal. It is clear, however, that there exists a dichotomy in parenting styles, even though its exact nature is only now being glimpsed.

To some extent, for example, the tendency toward sexist parenting can be dampened. Certainly the amount of attention many young fathers today lavish on their children far exceeds the amount their fathers once lavished on them. To grasp the difference, one need only go to a park on a Sunday afternoon. With more and more mothers working outside the home, fathers have had to pitch in, and many of them seem to do so willingly and well. Increasingly, it seems that fathering is becoming part of normal male behavior. Still, parenting remains an asymmetric endeavor, with women doing far more than men. We suspect that such asymmetries will always be present, but we hope that an awareness of sex differences in caretaking may help couples avoid some of the tensions and fingerpointing that arise as a consequence. Although such awareness will not, of itself, make it easy to be a parent, we hope that it will at least make it easier for well-meaning men and women to make sense of the sex differences that distinguish moms from dads.

CHAPTER 6
Childhood

THE SMALL GIRL learns that she is a female and that if she simply waits, she will some day be a mother; the small boy learns that he is a male and that if he is successful in manly deeds some day he will be a man.
— Margaret Mead, *Male and Female*

*E*veryone knows that little girls are not really made of sugar and spice and everything nice; nor are little boys made of snakes and snails and puppy-dog tails. Most people also know—without the benefit of fancy theories—that boys and girls exhibit behavioral differences, many of which are apparent at a very early age. From that magic moment when—in an instantaneous clap of engendering thunder—a ripe egg is fertilized by either an X- or a Y-bearing sperm, maleness or femaleness grows and develops intermittently, faster at certain times, slower at others, lying almost dormant for a while and then exploding briefly in a dazzling flood of hormones. In most individuals, each new stage of development follows a biological schedule that approximates the timetable of all other human beings, with certain parts rigidly fixed and others remarkably flexible.

Because some sex differences do not appear in young children but reveal themselves later, usually in adolescence, many observers argue that they are culturally produced, the result of accumulated social experiences, traditions, and biases inculcated during upbringing. Although such claims have merit, genes are an undeniable force in the unfolding of our sexual selves. For example, certain traits, which we describe in chapter 8, are established early in a child's embryonic development—laid down in patterns within the fetal brain—but not activated until the onset of puberty, when various hormones are released.

In fact, fascinating evidence has come to light suggesting that the preferences of little girls for babies and dolls and of little boys for footballs and trucks relate to hormones secreted while the child is still in the mother's womb. Subsequent exposure to dolls or trucks reinforces these differences but does not create them. Of course, other male and female traits, such as genitalia and body shape, are completely determined by biology. Still other sex differences, such as the way a child interacts with members of the opposite sex, reflect the combined influence of genes and experience. In this chapter, we examine both biological and environmental influences on children and show that neither operates independently of the other. In bringing these issues to light, we hope to give parents new insights into their relationships with their sons and daughters, as well as an understanding of what makes boys and girls so stubbornly different and why sex distinctions show up en route, before womanhood and manhood are fully attained.

Roots of Childhood Differences

Within the context of human evolution, it makes sense that boys, who are eventually to become men, should be biologically primed for a world in which those who are physically vigorous and sexually competitive will excel, whereas girls, who will become women, should be biologically primed for a world in which choosiness and nurturance are rewarded.

Renowned anthropologist Margaret Mead put a decidedly evolutionary spin on the essentials of male–female differences when she wrote:

> There is a long, long road between the lusty, exhibitionistic self-confidence of the five-year-old and the man who can win and keep a woman in a world filled with other men. . . . the little girl meets no such challenge. . . . Upon the initial uncertainty of her final maternal role is built a rising curve of sureness, which is finally crowned—in primitive and simple societies, in which every woman marries—with childbearing.

In accordance with evolutionary goals set long ago, boys and girls would be expected to express behaviors that help steer them on the path to becoming reproductively successful men and women. But human beings are also sentient animals, strongly influenced by the environments in which they are nurtured. Children are especially susceptible to environmental influences; they are disciplined and drilled to various degrees, molded to meet the expectations of their parents and the culture in which they live. Nonetheless, as any parent will attest, children are also born with their own distinctive personalities, which reveal themselves almost at birth.

How *can* one determine where genetic influences end and environmental ones begin? One way, of course, is through experimentation. Nearly four hundred years ago, James I, King of England (for whom the King James Bible is named), is said to have arranged for some babies to be reared in total isolation, piously hoping they would spontaneously speak what he took to be God's natural language, Hebrew. The result: not a Hebrew speaker among them.

Imagine a similar experiment to study the emergence of sex differences. Would babies raised in isolation until maturity display distinctive, if somewhat aberrant, male and female traits? Our guess is that they would. But there is no way to be sure. No one would permit children to be reared without social contact and thus, social influences. So we are left with a "thought experiment." Still, one can imagine a scenario in which a number of infant boys and girls are isolated at birth from their parents and from all social interaction. Or perhaps they are simply raised so that each child is treated identically, with no differences between boys and girls when it comes to experiences, opportunities, and expectations. They eat the same food, hear the same words, see the same sights, play with the same toys, and so on.

Under these conditions, would boys and girls grow up to be the same or different? Almost certainly they would grow up different. Recent studies of hormonal and genetic influences (including what was gleaned from the early kibbutz experiences) give an idea of *how* they would be different. Girls, we are confident, would generally be less competitive and aggressive than boys, more disposed to cooperation, more inclined to nurturance, and as they grew older they would become more interested in babies and other young children. Boys would generally be more physically competitive with one another, more adventurous, less cooperative, and less nurturing.

Evidence for biological influences on children's behavior comes, once again, from animal studies. Experiments have been conducted on rhesus monkeys, for example, that no ethically minded scientific review panel would ever permit for humans. As in the experiment reputedly undertaken by King James, these animals were reared in complete isolation. Within a few weeks, the young males showed significantly more aggression, as measured by threat behavior, than did females. Reared without social influences and without mothers, they knew nothing of social expectations or cultural traditions, only the dictates of their biology.

Interesting observations have also been made of bonnet macaques and squirrel monkeys. Among these primates, sex differences appear within a month or so after infants wander away from their mothers—in other words, *after* the most intense social influences have been severed. Moreover, the differences that develop are consistent and predictable: infant male monkeys are more likely to approach novel, complex, and arousing stimuli, whereas females are more likely to show wariness or fear. Such gender-specific behaviors are precisely consistent with evolutionary expectations.

Additional evidence that boy–girl differences of this sort are rooted in biology comes from elephant seals. Among these animals, large size contributes ultimately to male reproductive success, a fact that explains why the males are so truly elephantine as well as why male–male competition begins early, during nursing. For an aspiring harem master, a few extra pounds can make the difference between success and failure.

The more milk a pup consumes during the nursing stage, the greater his head start will be in the competitive fray to become harem master.

Burney J. LeBoeuf, an elephant seal specialist at the University of California at Santa Cruz, calls the most successful pups (those that achieve nearly twice the size of their peers) "super-weaners." It turns out that super-weaners sneak-suckle—that is, they steal milk from nursing females other than their mothers. Female pups, which are likely to reproduce as adults regardless of their size, do not sneak-suckle. For them, the dangers of stealing milk far outweigh the advantages: nursing mothers will bite and sometimes kill a young sneak-suckler. But if a male persists in his search, every once in a while he hits pay dirt and will find a female whose pup has just died, whereupon he takes the place of her deceased offspring. In this way, the little thief becomes a "double-mother-sucker," a lucky guy with access to two lactating females. Double-mother-suckers quickly become super-weaners, and in all likelihood they will become large and aggressive harem masters when they mature.

The elephant seal model helps illuminate the behavior of human beings by illustrating, again, that aggressiveness, competitiveness, and a degree of risk taking pay off for males. Not surprisingly, little boys are generally more troublesome, more feisty and more risk taking than little girls . . . not just like elephant seals but enough like them to send us a message.

Girls Being Girls and Boys Being Boys

Another way to make sense of the origins of sex differences in human beings is to compare very young children, presumably before culture and learning have had a chance to make boys, boys, and girls, girls.

Ethological studies on humans suggest that a kind of female coyness typically appears early in life . . . in girls. By age one or two, little girls—from Tasmania to Timbuktu—hide their eyes from strangers in a way that is described as almost flirtatious. This coyness occurs even in girls who were born blind and therefore could not have learned such behavior by imitating others. Comparable behavior is rarely seen among little boys.

Additional studies show that girls—even as infants—are drawn more to people than boys are. For example, when three-month-old babies were shown photographs of human faces and also drawings in which facial features were distorted, girls much preferred the real photographs, whereas boys looked equally at both. Similarly, girl babies in their cribs are especially inclined to stare at images of human faces, whereas infant boys are likely to find inanimate objects every bit as attractive. When older children were asked to look through a device that provides one eye with the image of an object (house, car, fire hydrant) and the other eye with a picture of a human being, most boys reported seeing objects and most girls focused, literally, on the people. Interestingly, this difference persists into adulthood: when shown images of people as well as of things, men tend to remember the things and women tend to remember the people.

When faced with adversity, such as hunger or cold, boy babies are likely to be fussier and harder to calm than girls. As a result, mothers typically spend more time attending to their sons' needs. It may be that boys are not more irritable, just more vulnerable to stress. Or maybe parenting styles tend to reinforce stereotypical behaviors, with parents more likely to stimulate and attend to boys than to girls.

But when newborn babies are well fed, rested, and dry, no differences in boys' and girls' behavior are noted. A likely possibility, therefore, is that when adverse conditions arise, females are more resilient— or perhaps more patient—than boys.

Other differences stand out in the behavior of young children. As early as three years of age, for example, boys react more vigorously to aggressive provocations from other boys; girls are more likely to ignore such forays, complain to someone in authority, or walk away. Psychological testing also reveals consistent differences: specifically, boys have greater mathematical and visual-spatial ability than girls, whereas girls have greater verbal ability, and by and large boys are more aggressive than girls.

Some people would nonetheless argue that such differences are merely stereotypes that persist because of parenting styles and social pressures, not because boys and girls are, at heart, fundamentally different. Without question, researchers interested in male–female differences among children face a daunting task. One way to pinpoint the origins of sex differences—in addition to studying the behaviors and aptitudes of very young children—is to examine societies in which boys and girls are treated as similarly as possible to see whether their behavior diverges and, if so, in what direction and by how much. One such society is the !Kung San of the Kalahari Desert. Here, gender roles are imposed barely, if at all, on young children. Boys and girls are both treated permissively, with great indulgence, and—as far as researchers can determine—indistinguishably.

Even among adult !Kung San, sexual roles are indistinct; men sometimes gather mongongo nuts and do "women's work" such as building huts, and women occasionally hunt, although they do not chase large game. Yet gender differences stubbornly show up, notably with boys wandering away from home, thus imitating the men, and girls staying nearby, as do the women.

More striking are the differences that emerge when the !Kung San settle into villages and become farmers. In this domestic setting, sex roles become established earlier in life and are more rigidly defined. Girls stay nearer to the settlement, where they care for young children and help with household chores, while boys tend herds of animals and drive monkeys and goats from the family gardens. Under these conditions, the sex roles of adult !Kung San also become more distinct and separate than they are among the remaining bands of hunter-gatherers. Perhaps the transition to farming mimics the prehistoric transition of human beings from a nomadic to a settled agricultural life. In any event, sex differences can be discerned even among young children, a precursor, perhaps—if not precisely, then in a general way—of the more defined differences seen in every settled human society.

Children at Play

All over the world, children play, whether with pebbles and sticks, water and sand, or paints, dolls, balls, and computers. Whatever the medium, the message remains pretty much the same: boys and girls play differently. Studies in the United States, for example, show that boys spontaneously draw pictures of monsters, adventures, war and other physical conflict, whereas girls draw cheerful, inanimate scenes (rainbows, houses, flowers) or images of helpful, cooperative interaction: a family going shopping, children playing on a swing, smiling animals, and the like. When asked to tell stories, girls are more likely to describe personal relationships, especially involving families. Boys are more likely to describe violence and various themes of destruction in which objects or other nonhuman forces (earthquakes, storms, scary animals) have center stage.

Indeed, children's play seems neatly divided into two camps: the boys' camp, which is known for physical activity, war games, and general jostling, and the girls' camp, which is considerably more cooperative, egalitarian, and nurturing.

In 1993, sociologist Barrie Thorne of the University of California at Berkeley summarized some typical differences between boys' and girls' play.

> Boys' groups are larger, and girls' groups are smaller (buddies versus best friends); boys play more often in public, and girls in more private places; boys engage in more rough-and-tumble play, physical fighting, and overt physical conflict than do girls; boys play more organized team sports, and girls engage in more turn-taking play; within same-gender groups, boys continually maintain and display hierarchies, while girls organize themselves into shifting alliances.

In other words, girls' games are more likely to involve taking turns, such as jump rope, hopscotch, and hand clapping. Victory is more ambiguous, with girls competing mostly for applause and admiration from observers. Boys participate more in contests that have clear winners and losers. Overall, it seems clear that girls more often mimic human relationships; boys, war. We would argue that over the course of human evolution, the expression of distinct behaviors in each case has helped boys and girls prepare for their later roles in life—those of competitor and mother.

Similarly, young girls are more likely to use adult styles of persuasion (appealing to reason and fairness), whereas young boys are more likely to rely on force or the threat of force. In all cultures, as far as we know, boys are more peer oriented yet at the same time competitive with their peers. To gain the approval and admiration of their fellows, it is often necessary to be at least a little bit bad. By contrast, girls are more likely to gain approval by cooperating, not only with friends but also with adult authority. Indeed, their desire to please adults helps explain why girls are far more likely than boys to be teachers' pets in school.

The following experiment supports the tendency of girls to seek the approbation of adults. When eleven-year-old boys and girls were asked to solve a complex problem involving a sequence of colored buttons, the boys worked longer when they were alone; but the girls were more persistent when the experimenter was present. This suggests that boys are motivated to achieve mastery for its own sake, whereas girls are motivated more by their social environment.

There is nothing about a child—not age, intelligence, race, and so forth—that is more likely to influence who he or she plays with than whether the child is a boy or a girl. From about age two to age six or seven, boys withdraw from girls, who they find boring, and girls avoid boys, who they find boorish. This preference in play partners holds cross-culturally: around the world, boys almost always choose to play with boys and girls, with girls. Of course, such segregation helps to reinforce and exaggerate whatever gender gap already exists.

Sex roles typically become more defined as children grow older, although boys remain more rigid in their play than girls. In her study of differences in boys' and girls' play, Barrie Thorne describes Jessie, a remarkably talented fifth-grade athlete who carved a niche for herself in the boys' culture. Such instances are notable, even fascinating, but viewed fairly, they do more to emphasize the usual backdrop of male–female differences than to overthrow the generalization that such differences exist. For every gender-bending Jessie, there are dozens of girls who do girlish things if not exclusively, then predominantly and boys who wouldn't consider doing anything but boyish things.

Sociologist Janet Lever of Northwestern University once spent a year observing and interviewing nearly two hundred fifth-graders in an attempt to delineate differences in their play. Her findings are nearly identical to those reported by Thorne. In addition, Lever calculated that 65 percent of boys' games were competitive, compared with only 35 percent of girls' games. Only rarely did Lever see mixed-sex play groups develop. When such groups did form, they almost invariably produced conflict because of the differing styles of play sought by the children.

Descriptions of children's play by educational psychologists Evelyn Pitcher of Tufts University and Lynn Schultz of Old Dominion University, though far from earth shattering, nicely demonstrate how boys and girls in exactly the same setting, using exactly the same equipment, construct markedly different play scenarios. Pitcher and Schultz describe watching two five-year-old boys play with wooden blocks. The boys transformed the blocks into a house and then a cave with monsters and rockets. One boy tries to assume dominance over the other, telling him what to do and not to do, but was ignored. As the boys' play continued, arguments developed about who took whose blocks, followed by a make-believe lightning storm, after which a supergun finally blew up the whole structure.

By contrast, when two five-year-old girls played with the same blocks, the structure became a home and the two girls assumed the roles of mother and daughter. Unlike that of the boys, the girls' play was bound to the reality of everyday life, with emphasis on home issues and interpersonal relations. During the game, there were shopping trips and an interruption to complain about the behavior of another girl, with "Mom" eventually granting permission for a trip to the zoo and for piano lessons while urging her "child" not to fight with her sister.

Pitcher and Schultz then observed another set of five-year-olds playing on a two-story wooden structure equipped with telephones, carpentry tools, tables, dishes, Tinkertoys, and a heavy rope connecting the upper and lower levels:

> First three boys came to play in the structure. One boy seized a hammer and began to attach pieces of cardboard to the rails, announcing, "I'm making a police station. I'm a construction worker." Another boy shouted, "Hey, it's the lunch break. I'm the boss. Don't forget I have machine guns." The boys moved rapidly to the tinker toys, put them together in long wands, and shot at one another; one boy announced, "We're in a space ship." Then another boy jumped to the lower level, and attached his tinker toy to the rope. The boys adopted deep, commanding voices as they organized the descent or ascent of the tinker toys. "O.K. pull 'er up." "O.K. throw it down."

When the boys eventually left, their place was taken by three girls:

> At first they spent considerable time arguing who would control the telephone. One girl finally got the phone and, with pauses apparently geared to accommodating the imagined party in the conversation, said, "My sister's gonna go to the school . . . She takes ballet lessons." The other two girls arranged dishes on the table,

put tinker toys on the plates, and said, "We have chopsticks for our Chinese food." Then one girl left the table, went to the lower level, picked up the rope, and held it to her ear, as though listening to a telephone. "I'm cooking breakfast. See you later," she announced before she skipped away. The other two girls followed her to another area.

For the boys, the play structure was a *police station* and they were *construction workers*. Then it became a *spaceship* and they were *warriors*, shooting one another. For the girls, the same play structure was a *house*; in it, they *prepared food* and *cared for one another*. The Tinkertoys—*ray guns* minutes before in the boys' hands—were transformed into *telephones* and *chopsticks*.

Reflecting on such differences, one specialist conjectured:

> If some mad sociologist should ever settle a thousand little boys in a compound and give them dolls to play with and give footballs to a thousand girls in another compound, I feel certain that within a few days a small minority of the girls would be kicking and throwing the footballs around, while the majority would be cuddling their footballs and scolding them for being naughty. And I'd bet . . . that 60% of the boys would have dismembered their dolls to use the limbs and torsos for batting the heads about the compound; and the 10% who went in for cuddling would have had their dolls stolen for dismemberment by the majority.

These observations uphold what seems to be a universal truth: nurturance is virtually absent in boy–boy groups but is commonplace among young girls. And although physical violence is undeniable in the play of young boys, it rarely materializes in the play of girls. In general, boys are more risk taking than girls, both in experimental settings and in their real lives. Their play also involves more testing of themselves against one another.

By about five years of age, boys are beginning to band together in large groups, whereas girls tend to remain in twosomes. And large groups, in turn, are more likely to lead to aggression; small groups, to intimacy. Psychologists have found—time after time and in every human group known—that rough-and-tumble play, something girls generally avoid, is almost synonymous with being a boy.

Beatrice Whiting and Carolyn Edwards, Harvard University child psychologists who conducted pioneering cross-cultural studies of child development, also found that boys are more confrontational than girls, especially with their parents. "We must ask whether boys elicit more dominance conflicts with their mothers and other social partners because they are genetically prepared to be more active and goal-oriented than are girls," write Whiting and Edwards. "Are boys found further from home because they take more risks in terms of physical activity than girls do? Are girls genetically prepared to be more responsive to other human beings, more interested in infants than are boys?"

These questions were revolutionary when posed in 1988, in part because Whiting and Edwards, like so many of their fellow social scientists, had been wedded to the idea that biology and genetics have no significant role when it comes to male–female differences. These and similar notions are now being toppled.

Treating Sons and Daughters Differently

"Man is a creature who lives not upon bread alone, but principally by catchwords," wrote Robert Louis Stevenson in *Virginibus Puerisque* (1881). "And the little rift between the sexes is astonishingly widened by simply teaching one set of catchwords to the girls and another to the boys."

For all our talk about biology, social influences obviously have a powerful impact on a child's gender identity. According to Harvard University psychologist Lawrence Kohlberg, children learn their gender by around two or three years of age, identifying that of others within the next year or two. Thereafter, the child's sense of gender deepens as a result of observing, identifying with, and imitating the behavior of parents, teachers, high-profile celebrities, heroic figures, and friends.

Parents may or may not intentionally teach their children to behave in stereotypical ways, although some parents certainly are heard to reprimand their children with such phrases as: "You're a girl; why don't you act like one?" Equally powerful may be the subtle, unspoken effect of societal expectations, including the presence of older individuals acting as role models. In many families, children are expected to act according to their gender and are rewarded for doing so, as well as punished if they don't.

In one interesting study, researchers sought to determine whether the sex of a child affected people's perceptions of his or her behavior. Toddlers from eighteen to twenty months of age were exposed to the same stimulus, designed to produce surprise. Observers, who did not know the actual sex of the child but had been told, randomly, that the child was either male or female, tended to characterize the child's response as anger if they thought the toddler was a boy and as fear if they thought the toddler was a girl. Apparently, boys are expected to be more angry and girls, more fearful. It is only reasonable that to some degree, children learn to meet these expectations.

Even among animals, young males and females are treated differently. Rhesus monkey mothers hold their daughters closer than they hold their sons and are more willing to allow their sons to wander during play. Sons are also weaned at a younger age and are more likely to be disciplined by their mothers. Such differential treatment undoubtedly helps explain why male and female rhesus monkeys grow up to be different. Whatever the reason for the parents' behavior, it seems clear that monkey mothers recognize the maleness and femaleness of their offspring, with an element of biological wisdom probably reflected in their parenting.

In our own species, parents have a tendency to differentiate between sons and daughters within twenty-four hours after birth. Infant girls are described as softer, smaller, less attentive to stimuli and finer featured than infant boys, whether they are in fact or not. In addition, parents tend to respond more emotionally to daughters and impute greater emotionality to them. We also find it interesting that parents will outfit a girl in almost any color, but most wouldn't dream of dressing a boy in pink.

As a child grows, parents continue to treat boys and girls differently. Mothers talk more to their daughters; in particular they ask their daughters more questions and are more likely to repeat what the children say. In contrast, little boys are encouraged, early on, to be the strong, silent type. Girls, in short, are rewarded for being feminine (which generally includes being more verbal); boys, for being masculine (which generally includes being more physical as well as more numerical). The result is something of a self-fulfilling prophecy or a chicken-and-egg problem: how can the influences of biology and experience be separated? To some extent, they cannot.

The Fulani people of Africa provide insights into how cultural expectations can sharpen naturally occurring differences between males and females. The Fulani—in contrast with the relatively peaceful and sexually egalitarian !Kung San—are notably aggressive and sexist. One anthropologist describes the typical experiences of young Fulani boys:

> At about six years of age the boys begin daily herding with their older brothers or fathers. At this time they are encouraged to begin to display aggressive dominance towards the mature bulls and oxen. . . . They are obliged to discipline these animals by charging them or hitting them with herding sticks. Boys who refuse to beat cattle on instruction are usually considered cowards, threatened, and even beaten if they still refuse.

It is small wonder that Fulani men are aggressive and often violent, highly competitive, and ill-tempered. This example, however, does not prove that upbringing fully accounts for behavior. One could as well argue that Fulani boys are taught to be belligerent—and take to it readily—because evolution has favored males who are more aggressive than their peers and thus are more likely to survive and reproduce. As ever, the truth almost certainly involves both biology *and* experience.

If Fulani parents sharpen naturally occurring boy–girl differences, other parents often mute some of those differences. Most North American parents, for example, claim that they strive to make their children—whether boys or girls—equally obedient, neat, and responsible and that they try to discourage aggressive, hurtful, or selfish behavior. But differences persist. For example, boys are consistently punished more, largely because they are more likely to transgress. Advocates of the "learning is all-important" school claim that this is a self-fulfilling behavior: because they are punished more, boys grow up to be more punishing. But such a claim ignores two crucial factors. First, boys are punished more *because* they aggress and transgress more. And second, the boys' misbehavior develops *despite* parental attempts to prevent it.

No one doubts, however, that parents exert a powerful influence on their children's behavior—through discipline, expectations, and role modeling. Still, in some cases, children find it remarkably easy to follow the parental lead, whereas in others, they find it surprisingly difficult, often to the frustration of everyone. Parents who push their Barbie-obsessed daughter to go outside and get dirty, or who press their football-crazed sons to play piano are likely to be exasperated, if not defeated, in the end.

Also revealing are the toys parents select for their children. When researchers examined the contents of firstborn children's bedrooms in upper-middle-class homes, they found that boys' rooms had a greater variety of objects and many more toy animals, aggressive figures, and, things related to space, energy, and science (rocket ships, magnets, puzzles), whereas girls' rooms had more ruffles and lace, floral designs, and of course, dolls. When dolls were found in boys' rooms, they were inevitably figures of fantasy heroes—cowboys, spacemen, robots—and almost never of babies or women.

To some degree, toy choice reflects the preferences of parents, but part of their reason for giving different toys to girls and boys is to meet the children's desires. Had we, for instance, given our daughters baseball bats instead of dolls for their early birthdays, we would have been drying a lot of resentful tears. In our view, adults generally are not a foreign occupying power, pushing their potentially androgynous children to be rigidly sex specific. Rather, in most cases, parents provide experiences appropriate to what their children actually want.

Often, adults may be unaware that they are helping to reinforce sexual stereotypes. W. S. Barnes of the Graduate School of Education at Harvard University studied the ways in which parents describe their offspring. He found that when there are only two children, parents show a strong tendency to describe them as opposites: if Suzie is intellectual and cautious, then Sarah is emotional and impulsive; if Michael's disposition is sunny, then Peter is seen as dark and brooding; and so on. We suspect that the tendency to paint children as opposites intensifies when one child is a boy and the other is a girl, as in "Connor's my wild one, but Emily's a little angel." We suspect that most human beings— like the parents in a two-child family—are psychologically attuned to look for and find oppositional distinctions between the sexes and then, perhaps, to amplify them.

Ethologists are increasingly convinced that living things are genetically predisposed to what is called "biased" or "prepared" learning, that is, they differ in what they bring to a learning experience and, hence, in what they take away from it. It seems very likely that human beings are similarly predisposed. The issue is not simply what we have been taught but also what we are eager to learn. Studies now show that boys and girls are prepared to learn somewhat different things. Give them the same experience and they will probably learn different things from it. Give them a choice and they will probably choose to have different experiences in the first place. Consider baby-sitting.

Interestingly, when boys care for younger siblings, they often become distracted in much the same way fathers do (as discussed in chapter 5). The following observations, made by Beatrice Whiting and Carolyn Edwards in Juxtlahuaca, a small Mexican village, are typical:

> Jubenal, a seven-year-old, and the eldest of four, was the only boy in our sample whose mother expected him to care for his siblings frequently. Although Jubenal appeared to be quite competently nurturant when necessary, in most observations when he was in charge he climbed trees, played games, called to his friends, and paid little or no attention to his siblings unless they cried or were in an obvious state of need. Unlike the typical Juxtlahuacan girl, who was most often observed combining child care with chores or casual sociability and gossip, Jubenal always tried to combine child care with active play with peers.

If girls around the world are more likely than boys to be involved in domestic work (food preparation, housecleaning, child care)—and they are—is the difference a consequence of biology or of culture? Probably both. Perhaps girls are assigned more child care than boys are because mothers want to train their daughters to be mothers themselves. Or perhaps boys, who are more predisposed to physical activity, are simply less suited to caring for young children, a task that typically requires them to stay put. Alternatively, mothers may request more domestic help from girls because girls are more amenable to parental demands generally. (Boys are more likely to resist parental rules and authority and girls are typically more docile and cooperative.) Or perhaps girls are more amenable to domestic chores because they are seeking to establish their own gender identity, and imitating their mothers in the process. In the end, we suspect that all these explanations are true to varying degrees, but we also believe that girls are intrinsically more interested than boys in babies and domesticity, simply by virtue of being girls.

However, no serious analysis of sex and gender differences, even one as avowedly biological and evolutionary as the one presented here, can deny the role of socialization in helping to make boys into men and girls into women. Psychologists Pitcher and Schultz offer a useful summary: "Through their play behavior, children steadily incorporate the gender role initiated by their biology, demanded by their psyche, understood by their mind, and supported by their culture."

When Parents Favor One Sex

In an ideal world, every child is equally loved. However, one of the tragedies of existence is that not all of them are. Under some conditions, males are valued more; under others, females. Again, there are no hard and fast rules, but it seems that evolution has had a strong hand in influencing whether parents prefer to have offspring of one sex or the other.

Some twenty years ago, Robert Trivers and his associate Dan Willard theorized that among polygynous species—those in which successful males receive a large evolutionary payoff and unsuccessful males end up with nothing—healthy successful parents should be inclined to invest preferentially in sons. Because those sons would be genetically well endowed (coming from strong, healthy parents) and nutritionally as well as socially advantaged, they, like their fathers, would be likely to acquire a harem and thus to repay their parents' investment in them. (Note: "Repayment," in such cases, occurs via enhanced reproductive success of the offspring and, therefore, of their parents.) Trivers and Willard then reasoned that parents who are somewhat less healthy than most or less successful socially should invest preferentially in females. The evolutionary rationale is that in a polygynous society, daughters are the more conservative option, likely to reproduce even if they are not especially healthy or attractive, whereas low-ranking sons are liable to be reproductive failures, outcompeted by other males. The prediction has since been validated many times over. Biologists, for example, have documented that among several species of polygynous seals, healthy females produce more male offspring; those that are less healthy produce more females.

Human beings, with their polygynous tendencies, invest differentialy in their offspring, although their preferences are mediated by cultural tradition rather than reproductive physiology. When Sonoma State University anthropologist Mildred Dickemann examined traditional societies in India and China, as well as historical records from medieval Europe, her findings were striking: in all these cases, upper-class families invested more in their sons and discriminated against their daughters; in fact, female infanticide was distressingly common.

Moreover, this behavior was consistent with each family's biological interest. Because upper-class sons in polygynous societies can expect to have many wives and therefore produce many grandchildren for their parents, they are favored over daughters. Although the daughters of upper-class couples also produce grandchildren, each daughter, having a limited reproductive capacity, produces fewer grandchildren on average than does each son. Thus, wealthy parents prefer to rear sons.

The opposite pattern holds true for lower-class families, whose sons have relatively bleak reproductive prospects as a result of their inability to compete with more prosperous males. But lower-class daughters are potentially able to "marry up" and thus have better reproductive prospects. True to evolutionary prediction, Dickemann found that lower-class families were more likely to expend their resources on daughters.

Although infanticide is not frequent in the United States, there is little doubt that parents often discriminate between their children depending on whether they are boys or girls. We would also guess that there is a biological pattern discernible in the amount families pay for weddings, athletic activities, and education for girls versus boys. In other words, we predict that wealthy families spend more money on their sons (for school, travel, automobiles, and the like), whereas poor families spend somewhat more money on their daughters (for clothing and other items to make them more attractive).

In her clinical practice, Judith commonly hears the laments of women who had to put themselves through college or graduate school while educational expenses for their brothers were paid in full. Although this pattern may be changing, it remains widespread. Judith's patient Elizabeth, for example, vacillated between outrage and depression when describing the inequities she faced while growing up. She was expected to help around the house, whereas her older brother Frank got special favors and expensive presents, such as a surfboard, simply because he was the oldest male. When her brother Jerry got into trouble at school and stole a car, the parents paid his bail, his attorney's fees, and his therapist bills without complaint, dismissing his behavior as high-spirited. But Elizabeth received few such favors. Even although she was an honor student, elected Phi Beta Kappa in college, she had to work her way through law school with no family support. Even now, family members don't seem as proud of Elizabeth, a prosperous attorney, as they are of Frank, who is a produce manager in the local supermarket.

The Whispering Within

There is much in American life that emphasizes gender differences among children, from blue and pink clothing to distinctions made by peers, parents, and others. Sometimes, the dichotomy may be unintentional, although nonetheless real, as when a teacher addresses a class with "Boys and girls" rather than "Students." Such tendencies are not necessarily bad. Given the social and biological realities of male–female differences, it would hardly pay for a parent to raise a gender-neutral or neutered child. People who seem indifferent to sexuality and love are not usually happier or more successful than their sexually aware counterparts. In our judgment, the key to raising children is to strive not for gender neutrality but for something both more feasible and more desirable: children who are comfortable with their sexuality. When such children become adults, they will most likely be "sexy" rather than "sexist."

But what if we did want to produce girls and boys who are essentially the same? How would we do it? Social science theory holds that we simply need to treat them identically, yet attempts to do just that have failed. A better strategy might be to treat them altogether differently, reversing their roles through various forms of punishment and incentive. Girls would thus be rewarded for being aggressive and boys would be punished for their aggression. Girls would be discouraged from playing with dolls and boys would be actively encouraged to do so.

Even then, we believe, the experiment would fail. We need only look to the parents who, attempting to rear their children free of sexual stereotyping, dutifully give a doll to their young son only to find him making swords out of the cardboard box and sending the doll off to war, or who give their daughter a carpentry set and then—to their chagrin— watch her tuck the hammer into bed at night.

It seems that despite the best efforts of parents and other caretakers to raise children in nonsexist ways, children constitute a fifth column: sexual counterrevolutionaries following the whisperings of their own evolutionary biology.

CHAPTER 7

Body

O BODY SWAYED TO MUSIC, O brightening glance,
How can we know the dancer from the dance?
— W. B. Yeats, *Among Schoolchildren*

*I*n Terry Gilliam's weird and funny movie *The Adventures of Baron von Munchausen*, a free-floating head—fond of poetry and philosophy—is horrified at the prospect of being reconnected to its body. "I have no time," it cries out indignantly, "for flatulence and orgasms."

By contrast, the rest of us have bodies, with all that implies for pleasure, pain, and, yes, flatulence and orgasms, too. As with other aspects of sex differences we have discussed so far, the respective anatomies of men and women show unmistakable signs of biology. Across the United States, curvaceous young women strut across television and movie screens and peer out from billboards and magazine racks. They vary in skin shade and hair and eye color, but without exception they are youthful and beautiful, and they are often buxom as well. Among male sex symbols, the emphasis is on the upper body: a straight torso and muscular shoulders and forearms. All of them—male and female—have healthy hair and good complexions. Although there are variations on these themes (plumpness is valued more in societies with few resources; slenderness, more in the United States, where healthy, wealthy people watch their weight and work out), similar images of what constitutes beauty and desirability can be found the world over. Beautiful women are expected to have prominent breasts, nicely developed hips, and relatively tiny waists. Men are supposed to be tall and strong; wide in the shoulders, and narrow in the hips. Apart from these idealized differences in body type, men and women are distinguished by other basic biological differences. Women have more subcutaneous fat; men have more muscle. Women have less body and facial hair than men and more high-pitched voices. And, of course, a man has a penis, a scrotum, and testicles whereas a woman has a clitoris, labia, a vagina, and ovaries.

In a few rare instances, maleness and femaleness are not clearly demarcated; for example, there are hermaphrodites—individuals born with female and male reproductive organs—as well as those whose sex is unclear at birth. But the overwhelming majority of people are either men or women, with bodies that are easily identifiable as one or the other. Individuals who blur the distinction between male and female are rare. Yet—perhaps because they are so rare—they both disconcert and fascinate us. (Recall Julia Sweeney's performance as the androgynous Pat on *Saturday Night Live*.) From toddlers on up, people are adept at identifying gender and become flummoxed when they cannot.

In many cases, people exaggerate their sexual characteristics, using artificial devices such as padded bras for women or, in Elizabethan times, codpieces for men to suggest they are better endowed than they are. In modern Japan, where a high-pitched voice is considered especially feminine, women often force themselves to speak in a manner that (to a Westerner) sounds falsetto and almost birdlike. Some men wear elevator shoes; some women dye their hair, don tight jeans, or subject their bodies to liposuction. When all is said and done, men and women expend enormous time and money exaggerating traits that make them appealing to the opposite sex.

Unlike the floating heads of Gilliam's imagination, bodies and minds are inseparable. When it comes to the workings of evolution and the doings of everyday life, it is often difficult—perhaps impossible—to discern the dancer from the dance.

Basic Differences

In virtually all species, males and females exhibit differences in metabolic rate, life span, body plan, eating habits, and so forth. As with humans, these differences are averages, not absolutes, but they are real. For example, Olympic medalist Gail Devers is an exceptionally fast runner; indeed, she can easily outrun the overwhelming majority of men, but she will never defeat the small coterie of elite male runners. The fastest men run faster than the fastest women, just as men in general run faster than women. Similarly, former college basketball star Rebecca Lobo, at six feet, four inches, stands taller than the great majority of men, and former United States Secretary of Labor Robert Reich measures somewhat less than five feet tall, making him shorter than most women. But such exceptions do not invalidate the rule: the average man is taller than the average woman.

From the Beginning

Right from the start, girls follow a slightly different trajectory from boys. At birth, girls are one to two weeks ahead of boys in their bone growth, despite the fact that newborn boys are generally heavier and longer. And girls typically experience a peak growth spurt of three to four inches a year at around age twelve, whereas boys don't reach peak growth until about age fifteen, when they add approximately four to five inches per year.

The amount of subcutaneous fat also differs in girls and boys. Beginning at birth, girls have slightly more fat than boys; as childhood proceeds, the difference increases. At adolescence, differences in body fat become even more pronounced. From the seventh to twelfth grades, for example, the percentage of body weight devoted to fat increases among girls from 21.8 to 24.0 percent, whereas among boys, the percentage drops from 16.1 to 14.0 percent. Even though boys tend to exercise more, a comparison of equally fit boys and girls shows that girls consistently have more body fat. Only the most athletic girls have body fat levels equal to that of the *average* boy.

The difference between boys' and girls' physical strength also increases as they age. In measurements of grip strength, for example, twelve-year-old boys and girls are virtually indistinguishable; by age seventeen, however, boys can squeeze an average of forty-five kilograms (ninety-eight pounds), whereas girls can squeeze an average of only twenty-seven kilograms (fifty-nine pounds).

Adult Differences

By adulthood, the physical differences between males and females have become slightly more pronounced. In the United States, the average man is five feet, nine inches tall; the average woman five feet, four inches. Men are also heavier: 165 pounds versus 135 pounds. Women's bones are less dense, and disproportionately smaller than men's. Women also have wider pelvises (to accommodate childbirth) and slightly shorter legs relative to length of the trunk. Women may be more heat tolerant than men; in any event, they don't sweat as readily.

In addition, patterns of fat distribution are sexually distinct. By adulthood, women have 25 percent body fat, whereas the average for men is only 15 percent. Women's fat tends to concentrate at the breasts and around the hips; men's, at the belly. Women also have somewhat more difficulty losing weight.

Many women would say that they are cursed by their fat, faced with an endless battle against saddlebag thighs and a protruding tummy. Not surprisingly, much of this frustration can be blamed on evolution, which has favored fat buildup in women. Women need a certain amount of body fat in order to have normal hormonal cycles; when body fat gets too low, ovulation and menstruation cease. As troubling as fat may be, some amount is necessary for fertility.

Pound for pound, men have more muscle. Muscle accounts for 23 percent of the body mass of the typical woman, whereas in men it accounts for about 40 percent. This difference can be traced in part to testosterone, an anabolic steroid that stimulates the development of muscle. So powerful is testosterone that—despite its serious side effects—both male and female bodybuilders are tempted to take synthetic forms to increase their muscle mass and boost their performance.

Generally, women possess about 65 percent the muscular strength of men, with the greatest disparity in the arm and shoulder muscles. Exercise might explain some of the difference. It is boys, by and large, who do most of the ball throwing, swinging from trees, and lifting. Similarly, men do more heavy lifting and throwing than women. Exercise might also explain why the legs show the least disparity in muscular strength: from an early age, girls and boys are equally likely to walk or run to get from place to place. Still, some differences in upper- and lower-body strength are clearly built in. One need only look at professional weight lifters (all serious male pumpers of iron can bench-press dramatically more than their equally serious female counterparts) or professional cyclers: a Greg LeMond will always outpedal a Rebecca Twigg.

Men also perform consistently better than women in competitions involving sudden outputs of strength or speed: sprint, discus throw, shot put, jumping. Men utilize oxygen more efficiently, having more hemoglobin per ounce of blood than women. Thus, they are somewhat more effective energy utilizers, at least in short bursts. It is easy to imagine how evolution favored upper-body strength among males: men who had strong arms and chests probably would have been better at spear throwing and other forms of hunting, as well as hand-to-hand combat. They would have been the ones to bring home the bacon and best their competitors and, as a result, win the affection of women.

Interestingly, though, in athletic events that require particular endurance—such as ultramarathons (50 or 100 miles) and swimming competitions that are measured in miles instead of meters—women tend to do as well as or better than men. Still, the only Olympic sports in which males and females compete directly against each other are equestrian: show jumping, eventing, and dressage require not only basic physical fitness and strength but also extraordinary tact, subtlety, calculation, courage, and balance. In these latter qualities, men and women are equal.

Evolution of the Human Physique

Among the relatively few species in the world that are monogamous, males and females tend to be equal in size, as males do not have to compete ferociously for access to females and thus have no need of excessive bulk. For example, common seals, leopard seals, and harp seals generally mate one-to-one, and the body sizes of males and females are about equal. Among polygynous species, however, males and females tend to differ drastically in size. This difference in body size, or dimorphism, between males and females can be startling: male fur seals and sea lions commonly weigh two or three times more than females. Elephant seals are highly polygynous, and males weigh as much as eight times more than females. In human terms, such dimorphism would mean that a 120-pound woman would mate with a 960-pound man!

The same pattern holds for primates. Among monogamous species such as the gibbons of South Asia or the marmosets of Central and South America, males and females are about the same size. But once again, when mating is polygynous, as among baboons or gorillas, males may be 22 percent taller and 80 percent heavier than females. If humans were as sexually dimorphic as baboons, wives would weigh only half as much as their husbands—not unheard of but certainly not the norm. Human beings are only mildly dimorphic, although evidence suggests that among our species sexual dimorphism has decreased over the past 2 million years or so. Do such findings suggest we were more polygynous in the distant past and are more monogamous now? We believe that they probably do.

Too Short, Too Tall . . . Just Right

University of Texas psychologist David Buss suggests—accurately, we believe—that through much of human prehistory there was an evolutionary payoff for women who preferred tall men. Larger, stronger men would have been likely to offer greater protection to the women associated with them and to have more success in obtaining food and other resources than those men who were smaller and weaker. Tall men also would have been likely to father sons who were themselves taller and stronger and, hence, more likely to be successful. Thus, tall men would have been desired by women, who would see them as possessors of both good genes and good resources.

Even today, women consistently express a preference for tall men. Of course, women value intelligence, personality, and various other traits but overall, taller men have a variety of advantages, many of them related to social dominance. Not only do women generally prefer taller men as potential mates, but men also are more likely to defer to taller men . . . which probably helps explain the preference of women. In the United States, for example, taller men are disproportionately represented among the ranks of top executives. Few American presidents have been less than six feet tall. In fact, one of the most persistent criticisms leveled against Ross Perot during his 1992 presidential campaign was that he was too short.

Because such "heightism" is so widespread, for men short stature can be an overwhelming impediment to social approval, self-esteem, and reproductive success. Judith feels sympathy for her patient George, a genuine genius, but who lacks the physical presence of the average American male. He is unusually short for a man, at five feet, two inches. Accustomed to being overlooked and discounted by men and women alike, George has recently taken refuge in on-line chat rooms, where he can strut his intellectual stuff, make friends, and influence people, with no one knowing his height. Not surprisingly, he also enjoys "cybersex": George attracts women with his wit and then engages in pleasurable "dirty talk." On several occasions, people have asked to meet him in person, but he always declines. "I suppose I identify with the hunchback of Notre Dame," he confides.

For most women, short stature carries far less stigma. It also doesn't impede romantic success; in fact, many men—including David—find petite women especially attractive. (In contrast, we doubt that most women—even those married to short men—would ever express a preference for diminutive men.) But women pay a price for being small; they are perpetually infantilized, seen more as children than as mature adults.

Judith, who is barely five feet tall, has had many experiences typical of petite women. Until her hair began to gray, she was regularly carded in restaurants when she ordered wine; once or twice a patient has patted her on the head; and she was often mistaken for a student while making hospital rounds during residency training. In retribution, she takes great delight in driving a Ford 350 pickup truck with "dualies," for the heady sense of power the truck confers.

Women who are unusually tall may find themselves stars on the basketball court or fashion models in New York, but on balance tall women do not enjoy pronounced social or economic advantages over medium-height or short women in the same way tall men prosper at the expense of their shorter competitors. Many, in fact, feel gangly and adopt a stooped posture, as if to minimize their height. And some have difficulty finding mates, since society expects that however tall a woman is, her husband or boyfriend should be taller yet.

With such emphasis on height, why are human beings not taller than they are? If height confers such an advantage on males, why isn't the average man something like seven feet tall? In fact, why are women not taller as well? It seems likely that women who married tall men would produce tall daughters as well as tall sons. Assuming that these daughters also selected mates taller than themselves, it would seem that they would produce even taller children in a potentially endless cycle of ever-increasing height. Although human beings are bigger now than they were thousands of years ago, they are not excessively so. Why is this the case?

It seems there are two explanations. First, plain, unvarnished size isn't always desirable; the bigger they are, conventional wisdom tells us, the harder they fall. As anthropologist Sarah Hrdy aptly points out, "The virtues of large size are not limitless. . . . Limitations to male size include availability of food and the restrictions of gravity. Orangutans are among the most arboreal of apes; yet, a fully grown male (weighing up to 165 pounds) may become so large that the forest canopy no longer supports his weight and he is forced to travel long distances by walking along the tangled, leech-infested forest floor." She also notes that the male orangutan must consume "vast quantities of unripe fruits and mature leaves, the junk foods left by more discriminating females," in order to support his great bulk. In contrast, the female, which may be half the size of the male, "can afford to be a picky eater, selecting the nutritious shoots of new leaves and the ripest fruits."

In short, if sexual considerations are set aside, bigger is not necessarily better. Among polygynous mammals, females generally tend to be closer to the ecological ideal in that they require less food and fewer resources overall: the larger body size of males means that they require more food to keep them going, even as infants. Male fur seal pups, for example, drink nearly one-third more milk than do females, and it appears that young elk males drink so much milk that they weaken their mothers: female elk that rear sons are more likely to die during the year after calving than are females that rear daughters. Even if they survive, these nutrient-deprived females are less likely to calve again the following year.

The second likely reason why men are not universally tall has to do with our essentially monogamous pattern of mating. Although we as a species have a penchant for polygyny, in reality human beings are only mildly polygynous and few become big-time harem masters. Monogamy is a great evolutionary leveler, since it increases the chances that most men, regardless of height, will have a shot at reproduction. Were we a rigidly polygynous species dominated by harem masters, men under six feet tall, for example, wouldn't have much of a chance.

It is interesting to note that in a broad range of animals, it is not always the more competitive sex that has the larger body. In most species of frog and bats, for example, as well as in some toads, snakes, insects, spiders, and fish that give birth to live young, females are larger than males. Among these animals, larger males are often more successful than smaller ones, but overall, large size benefits females more than males. Simply put, among nonmammals in particular, larger females often produce more or larger offspring or both, and thus it may pay for females to be bigger than males, although the males may still compete vigorously for access to them.

What Appeals to Men About Women

By adulthood, women have 40 percent more fat in the lower trunk than men. Many women will say that no matter how hard they diet or how much they exercise, they just can't seem to shed unwanted fat around their hips and thighs (a difficulty that accounts for the great popularity of liposuction).

As with most aspects of our sexual beings, such tendencies are rooted in biology, related to a woman's ultimate *evolutionary* goal: giving birth. The fact that women, rather than men, get pregnant and lactate, seems to explain why women have a higher proportion of body fat. Extra metabolic reserves, stored as adipose tissue, almost certainly contributed to the ability of our ancestral mothers to carry healthy babies to term and to provide milk for them afterward.

Even today, a minimum level of body fat is needed before an adolescent girl begins menstruating, and women whose fat reserves drop below a certain level—most notably serious athletes and anorexics—often experience interruptions in their menstrual cycles. When they stop their intense training or start eating normally—and, as a result, increase their fat supply—regular cycling returns. The correlation between female fat and fertility makes evolutionary sense, of course, because the developing fetus requires a lot of calories. Extra body fat represents insurance in the event of lean times.

No similar correlation has ever been demonstrated for males—that is, between body weight, fat level, and sperm production—although the latter waxes and wanes according to a man's body temperature and frequency of ejaculation. Couples struggling to become pregnant may be counseled to reduce their frequency of sexual intercourse or avoid drugs and alcohol in order to increase the man's viable sperm count, but men are not advised to become chubbier as a way to become more sexually potent.

So Feminine a Shape

Although men do not generally find overweight women especially attractive, they do show a clear preference for the curvaceous hourglass figure of a sexually mature woman. Geoffrey Chaucer, in *The Canterbury Tales*, had no doubt what made a woman attractive: "buttockes brode and breasts round and huge." In fact, many men would describe their ideal sex partner as looking like *Playboy*'s playmate of the month rather than one of the more emaciated fashion models. From an evolutionary perspective, such preferences make sense; they are the lingering whispers from our Pleistocene past that tell a man to choose a body type most likely to produce successful children.

Stephen Dedalus, James Joyce's young hero in *Portrait of the Artist as a Young Man*, came up with the following analysis while musing with his friends on the nature of female beauty (as perceived by men): "Every physical quality admired by men in women is [directly connected to] propagation of the species. . . . You admired the great flanks of Venus because you felt that she would bear you burly offspring and admired her great breasts because you felt that she would give good milk to her children and yours."

Human beings are unique among mammals in sporting prominent breasts when not lactating and also in making erotic use of them.

Maybe Stephen Dedalus was correct in surmising that well-developed female breasts let a man know that his offspring would have an adequate supply of milk. But the correlation between breast size and milk production is actually very low, mostly because the greatest proportion of a nonlactating woman's breast is occupied by fat, with virtually no glandular tissue. (Milk-producing glands develop during pregnancy.)

Because men are so fascinated by breasts, women sometimes become fixated on their own. Judith has counseled women whose breast size causes them much angst. One of her patients, for example, is Georgina, a slender, blond woman with clear skin, lovely blue eyes, an animated face, a sharp intellect, and small breasts. Despite her good looks, she has a truly phobic aversion to disrobing in anyone's presence, and she refuses to wear a bathing suit or a scoop neckline. Furthermore, fearing that any man who sees her undressed will either laugh at her or be repulsed, she will not date. She is considering refinancing her house to pay for breast augmentation surgery so she can feel like a "real woman."

Susanna, by contrast, is exceptionally buxom. By age twelve, she was too jiggly to ride a horse or play active sports. By the time she entered junior high school she was being teased viciously by other girls and occasionally groped by boys. By age sixteen, Susanna resolved to have breast reduction surgery. She was tired of the weight on her chest, tired of being teased, sick of lewd remarks and leering adults. Fortunately, Susanna's medical insurance covered the cost of the surgery because excessively large breasts can give rise to medical problems, including chronic back pain.

Perhaps long ago, men preferred women with relatively large breasts and hips because the latter indicated room for a baby to be born and the former, ability to provide milk afterward. In reality, of course, fatty breasts are no more a guarantee of subsequent milk production than fatty hips truly indicate a wide birth canal. In one way, therefore, breasts could be considered a case of successful but false advertising.

But men, for all their interest in breasts, might also have preferred women with comparatively small waists as a way of keeping women honest about what they had to offer. A small waist might suggest virginity, since a thicker-waisted woman might be pregnant with another man's child. In addition, as most women can attest, because childbearing tends to spread a previously youthful figure, a man who marries a narrow-waisted women is less likely to have chosen a mate who already has children, which may need his resources. Over time, if women with slender waists and large breasts reproduced more than women who lacked such attributes, the end result would be the sexually appealing hourglass female figure. Even though in modern times obstetricians can readily deliver babies by cesarean section, and infant formula can substitute for breast-feeding, the hourglass shape persists as an object of male desire—and, often, of female consternation.

Devendra Singh, a psychologist at the University of Texas, wanted to know exactly what makes a woman's shape appealing to men. He decided to measure the ratio of waist to hip measurements, which changes over time for each woman. At the prime of their reproductive years, healthy adult women typically have a waist–hip ratio ranging from 0.67 to 0.80, whereas the ratio for healthy men lies between 0.85 and 0.95. After menopause, women's waist–hip ratios approach those of men. It turns out that high levels of estrogen (actually, high estrogen–testosterone ratios) stimulate fat deposition around the hips and inhibit it in the abdominal region. Therefore, a low waist–hip ratio signals relatively high reproductive ability, whereas a high waist–hip ratio can be a sign of illness, pregnancy, or old age.

Not surprisingly, Singh found that men's idea of an attractive woman correlates with a low waist–hip ratio: 0.80 is preferred to 0.90, and 0.70 is preferred to 0.80. Examining images of *Playboy* centerfolds and beauty contest winners over the past thirty years, Singh found that the waist–hip ratio remained unchanged, at precisely 0.70, despite an overall decrease in the women's weight. (Again, plumpness is valued in societies in which resources are scarce; slenderness is prized in societies, in which wealthy, healthy people can afford to be slender.)

Health and Youthfulness

Beauty is . . . not so much as beauty does but, rather, as it is perceived. Some aspects of physical attractiveness vary according to social convention. For example, among certain societies, tattooed faces, elongated necks, or knocked-out front teeth are de rigueur. In others, shaved arms and legs, long legs, trim ankles, or dainty feet may be hallmarks of a sexy individual.

Other features men consistently look for in women and that women look for in men include regularity of features, quality of skin and hair, appropriateness of body size, and so forth: all traits that correspond to overall health. Recently, the importance of symmetry has come to light. When people are asked to evaluate the attractiveness of human faces that have been computer altered to reflect various degrees of symmetry, the results are clear-cut. The more centered the nose and mouth are and the more equally placed the eyes are, the more attractive the image is. Again, biology seems to have influenced our attitudes toward symmetry. It is well documented that disturbances during growth and development of the embryo due to inadequate nutrition, genetic anomalies, or other adverse conditions produce asymmetric features. The poet John Keats wrote that "beauty is truth, and truth, beauty." To some extent, this could be rewritten: "beauty is symmetry, and symmetry, beauty."

When it comes to a woman's sex appeal, however, nothing—not even symmetry—is in the same league as youth. Across all cultures, men consistently express a fondness for youthful women. Other animals show no such preference. A male dog, for example, is every bit as interested in an elderly bitch, provided she is in heat, as in a young one. Stallions do not discriminate in favor of young mares, nor do male chimpanzees limit their sexual interest to young females. In fact, primatologist Jane Goodall reported that the elderly female Flo was the most sexually sought after of the many chimpanzees she studied at Gombe Stream Reserve in Africa. Yet in every human culture, men are especially attracted to younger women.

Of course, even though men may be sexually attracted to younger women, many would rather have a lasting relationship with someone their own age than settle down (or try to) with someone who is much younger, albeit sexier. We would also bet that when it comes to sex, most men would prefer, say, Sophia Loren to any number of women half her age. Still, as a rule (for we are speaking in statistical generalities), the preference of men for younger women is undeniable.

When the female body is viewed through the objective lens of evolution, it appears that many female traits, regardless of a woman's age, are also characteristic of younger individuals, regardless of sex. Shorter stature is an obvious one, as are a higher-pitched voice and greater amounts of subcutaneous fat. Further, both women and children have significantly less body hair than do men, and women are generally more lightly pigmented as well (though skin tone in both sexes tends to darken with age). Moreover, women are widely considered attractive if their feet are dainty rather than large and their noses and chins are not prominent. It may well be that these traits—like the waist–hip ratios we considered earlier—provide information not only about youthfulness but also about estrogen–testosterone ratios. Thus, by following their sexual preferences, men appear to be furthering their biological interests.

This sexual fixation on youth doubtless has something to do with the human tendency to form long-term relationships. If human beings simply separated after coupling, men, like chimpanzees, would probably show equal enthusiasm for any sexually mature female who crossed their paths. But in choosing a lifelong sexual partner, it makes sense for a man, who remains more or less reproductively capable throughout his life, to choose a youthful mate with many reproductive years ahead of her. It is not adequate merely to say that men prefer young women because youthfulness is more sexually attractive. The question is, *why* is youthfulness more sexually attractive? Also, why is this correlation of youthfulness with sexual attractiveness more pronounced when the objects are female?

The universal male preference for youthful women can legitimately be scorned as sexist and certainly unfair. It is sexist because it applies more to one sex than to the other and unfair because women cannot choose to be beautiful or young or (in many cases) healthy, any more than men can choose to be broad shouldered, athletic, handsome, or (in many cases) successful. Women—like men—have every right to complain and to blame the cosmetics, advertising, and entertainment industries for presenting unrealistic goals and raising unreasonable expectations on the part of men and women alike. But skin-cream manufacturers, lipstick makers, and purveyors of diets and tummy tucks do not create the demand for their products so much as they cater to the preferences of their buyers, even if these commercial establishments exaggerate whatever inclinations already exist.

What Appeals to Women About Men

In most species, it is the males that are brightly plumed or adorned with weaponry, whereas the females are comparatively plain. But human beings, who lack gaudy anatomical features, are unusual: typically it is the women who primp and preen; struggle with lipstick, eyeliner, and mascara; adorn themselves with jewelry; and fill their closets with attractive clothes. Why are women so unusual among animals when it comes to beautification? The answer probably has to do with how much human fathers invest in their offspring relative to most other male mammals. Because of their high investment, men are limited in the number of women they can support. Thus, they are under pressure to choose the best possible mates, who, in turn, try to be as appealing as possible.

It is unclear to what extent women's choice has influenced the evolution of the male body. As we have stated previously, height and muscularity are attributes of dominant males. What about other male physical traits, such secondary sexual characteristics as greater amounts of facial and body hair, tendency to baldness, and lower-pitched voice? Are they attractive to women, and if so, why?

Broad shoulders are not simply decorative: they promise upper-body strength. But beards are more mysterious. They might not seem all that important, but beards are fascinating for several reasons. To begin with, facial hair is less prominent in some groups (Asians) than in others (Europeans). Not only do beards make their bearers seem more threatening, they nearly always make them appear older, which may be desirable if a man has a "baby face" and looks younger than he really is. As discussed earlier, a man's age often signals status and resources and is therefore something males may exaggerate, just as women may make themselves more appealing by emphasizing their youthfulness. Moreover, a beard can make a man appear more learned and thus desirable for his intellect. An added benefit is that beards help hide an unattractive chin or a flawed complexion, thus helping someone appear handsomer or healthier than he really is.

Men are renowned—sometimes infamous—for ogling the bodies of women, often "undressing them with their eyes." The truth is that women look, too, although they are generally less influenced by the shape and appearance of men's bodies than men are by the bodies of women. Many women will acknowledge that they are deeply aware of a man's eyes and, sometimes, his shoulders. Occasionally, they will admit to other physical considerations, such as a man's buttocks, hands, legs, or forearms. Obviously, male physique does matter, if not in as obviously a sexual way as it does for females. We cannot imagine, for example, that a woman—given the choice of partners for a one-night stand—would choose a paunchy, middle-aged accountant over a Chippendale dancer, no matter how kind and sensitive the accountant might be.

During the winter of 1995, when actor Keanu Reeves played Hamlet in Winnipeg, Canada, every performance was a sellout. This dramatic increase in theater attendance was reportedly due to a remarkable demand for tickets by women, who were said to have been inspired less by a sudden interest in Shakespeare than by the chance to see a sexy star strut the stage clad in revealing Elizabethan tights.

Still, it generally remains true that when it comes to sex appeal, men are judged for their bodies far less than women are. Henry Kissinger once remarked that power is the most potent aphrodisiac. Former U.S. Representative Patricia Schroeder spoke for the other side when she noted ruefully that for some reason, powerful middle-aged congresswomen don't seem to exert the same magnetic attraction for members of the opposite sex that equally powerful middle-aged congressmen do.

For Kissinger, as for other male politicians, power is measured by wealth, influence, and authority rather by than physical capacity. And as we have already seen, this type of power translates into control of resources, so females who mate with such males end up with real reproductive advantages.

When a woman marries, choosing her mate for better or worse, in sickness and in health, it is assumed that she is getting his genes, behavior, and resources all wrapped up together in a kind of romantic-reproductive "package deal." On the other hand, modern medicine in the form of sperm donation offers women the opportunity to separate those "goods." Whereas a woman selecting a live-in mate would be expected to place emphasis on good behavior and good resources, a woman choosing a sperm donor might be more concerned with "good genes," as indicated by overall health and physical characteristics, such as height.

In studies conducted in Canada and Norway, Joanna Scheib of the University of California at Davis found that when selecting hypothetical sperm donors, women indeed valued health and physical attributes (good genes), whereas when selecting hypothetical mates, they were especially concerned with other traits, notably character: whether a potential husband was kind, understanding, dependable, considerate, affectionate, and so on. It is also interesting that the women indicated a high level of concern for "character" even among sperm donors, although they also judged that character traits were unlikely to be "inheritable." One possible explanation is that evolution has long favored women who made fitness-enhancing mate choices based on all three goods combined—genes, behavior, and resources—whereas the opportunity to choose mates on the basis of genes alone is comparatively recent, and therefore women's judgment is refined. Another, equally likely, explanation is that traits such as character may in fact be heritable and that at some level, women know this.

The situation is complicated, however, since it seems likely that women have long had the opportunity—at least in theory, and occassionally in practice—to marry men who offer desirable behavior and resources and then choose their own sperm donors; that is, have extramarital affairs with men who appear to possess especially good genes. We know of a very beautiful "thirty-something" woman who was courted with great fanfare by her wealthy, middle-aged boss. Partially won over by his status as well as his lifestyle, she agreed to marry him, but when it came to having a child, she chose to become pregnant by a man other than her executive husband. She was flying to a friend's wedding in another state when by good fortune (her account), a man sat next to her who looked like, and in fact was, a well-known actor on holiday. They chatted for the entire flight, and then on impulse, the woman gave him the telephone number of her weekend destination. He called and invited her out for a drink; they spent a terrific weekend together and then parted forever. "Just think how lucky my baby is!" she crowed. "He will inherit [from her lover] the best looks on the planet, and [from her husband] a fortune to go with them!"

This woman claims to regret having deceived her husband and fears that someday he may suspect he is not her child's biological father. But she does not regret her actions entirely because she knows that her child is destined to have good genes and good resources too, and thus is headed for success.

Living Longer

In the most straightforward, biologically meaningful test of body strength—remaining healthier and living longer—women win, hands down. From Arabia to Zimbabwe, women live longer. In the United States, for example, the average life span for women is seventy-nine years; for men, seventy-two. In Japan, the United States, and Western Europe, where life spans are especially long because of high living standards that include good nutrition, high-quality medical care, and pro-active public health programs, the male–female gap is greater than in developing countries, where women are more likely to die during childbirth and from botched abortions. Some might say that differing social roles account for some of these differences in longevity, but the same seven-year difference has been found among American nuns and monks, who live similar lives and are generally isolated from the hurly-burly of modern life.

One explanation points to testosterone. A study of mentally retarded men who had been castrated early in life, and thus deprived of most testosterone, showed that they lived nearly fourteen years longer than mentally retarded men who had not been castrated. Nonetheless, the connection between mortality and testosterone may be only indirect. As discussed in the next chapter, testosterone exerts behavioral effects in addition to influencing the physiological processes of growth and metabolism. Moreover, hormones are unlikely to explain why during every decade of life, males die at a higher rate than females.

Actually, the difference in longevity begins before birth. Many more boys than girls are conceived (120 to 150 males for every 100 females), but by birth the ratio of boys to girls drops to about 105 to 100. In other words, males are far more likely to be spontaneously aborted than are females. In the birth process itself, boys are more susceptible to in-jury, perhaps because they are somewhat larger and have a more diffi-cult journey through the birth canal.

During infancy, the pattern persists: more boys than girls die, re-sulting in a ratio of male to female mortality of 1.27—that is, 127 baby boys for every 100 baby girls. Throughout life, and in relation to most ailments, this sex-skewed ratio persists, with more males than females dying in every age group. For example, the ratio for diabetes is 1.02 (that is, 102 males die of diabetes for every 100 females); for all cancers combined, the ratio is 1.51; for pneumonia and influenza, 1.77; for heart disease, 1.99; for cirrhosis of the liver, 2.16; for accidents, 2.93; for suicide, 3.33; for lung cancer, 3.43; and for homicide, 3.86. Some of these deaths are socially induced, at least in part. Men smoke more, drink more, and generally are greater risk takers and therefore more prone to accidents. Interestingly, once it became socially acceptable for women to smoke, their rates of lung cancer began to approach those of men.

Still, the ten most deadly diseases—from heart attack to pneumonia and cancer—strike men more than twice as often as women. Those diseases that predominate among women are often specific to the female body plan, such as breast cancer and bladder infections, or derive from hormonal differences, such as osteoporosis. An interesting exception involves certain ailments of the immune system such as lupus, multiple sclerosis, and myasthenia gravis, which are more common among women.

When it comes to psychiatric disorders, men are more likely to display a range of antisocial behaviors; they are also more likely to suffer from learning disabilities, attention deficit hyperactivity disorder, and autism. By contrast, women outnumber men in depression, anxiety disorders, and hypochondriasis. Both sexes are equally likely to suffer from manic-depressive illness, schizophrenia, and obsessive-compulsive disorder.

Women are more likely to suffer from depression, and they attempt suicide more often, typically by drug overdose, wrist laceration, and other self-destructive but nonlethal behaviors. In contrast, suicidal men often leap to their death from a bridge or building or put a bullet into their brains. Put another way, women's suicide attempts tend to be distress calls, whereas men's are more goal oriented. Once they make the decision to take their own lives, men are far more likely to succeed. Men with conduct disorders and drug or alcohol dependence are at especially high risk for suicide.

Overall, as the more risk-taking, more competitive sex, men are more likely to engage in flamboyant behavior, including its more pathological extremes. As we have emphasized many times, to be reproductively successful, men need to stand out from their competitors. Maybe in the efforts to do so, some go over the edge. Notably, this sex difference peaks in early childhood and late adolescence, precisely when risk taking among males of polygynous species is greatest. By no coincidence, automobile insurance is most expensive for unmarried males under age twenty-five.

Even so, no one knows why women are more disease resistant than men. A number of researchers believe that because women possess two X chromosomes, they gain some protection from whatever harmful genes may be present on the other X. (Not only do healthier genes on one chromosome tend to dominate and mask their less healthy counterparts on the other chromosome, but also the Y chromosome carries very few genes). Accordingly, men, who are XY, have no protective counterpart and thus are vulnerable to any deleterious mutations that may appear on their "unprotected X."

On balance, the differences between men and women are cause for celebration, not regret. People in the throes of passion do not consciously think about evolution in action, about the primordial pull of two bodies each drawn to the other in an ancient ritual of procreation. Yet when bodies encounter bodies, and their associated neurons begin firing, behind the creaking and groaning of ancient anatomical gears and the drip of hormones, the flashing or averting of eyes, the mutual displaying and assessing of anatomy no less than intentions . . . in all these things, there is the footprint of our biology.

CHAPTER 8

Brain

You, YOUR JOYS AND YOUR SORROWS, your memories
and your ambitions, your sense of personal identity and free
will, are in fact no more than the behavior of a vast assembly
of nerve cells and their associated molecules.
— Francis Crick,
The Astonishing Hypothesis

*T*he sexiest organ in the human body is also the
smartest, silliest, bravest, and most cowardly, the
part that can solve an equation, enjoy a movie, or
admire a sunset; indeed, it is the most important part of being
human. The organ in question is, of course, the brain, a three-
pound, electrically charged organ, criss-crossed with literally
billions of tiny, branching connections. It is the brain that dis-
tinguishes between the magic of romantic music and the bray-
ing of cacophony, that makes a touch thrilling, unnoticed, or
painful, that makes an odor evocative or revolting. Our brains
can make the commonplace exciting, and vice versa. There is

no question that the brain influences sex. What is less widely known, outside of research circles, is that sex influences the brain: a brain operates differently depending on whether it is ensconced in a male or a female body.

As with every other part of the human body, the brain—for all its glories—has been created for one reason: to promote the evolutionary success of whoever bears it. In a very real sense, therefore, we all have sex on the brain. It is not that we go around brimming with sexual plans and fantasies (although many of us do); rather, we behave in a manner designed by evolution to promote biological and social success. Accordingly, evolution has favored brains that differ depending on whether they belong to men or women.

How Genes and Hormones Influence Sex

Despite their wide array of physical and behavioral differences, the genetic difference between males and females is small. Of the twenty-three pairs of chromosomes in the human body, only the Y chromosome ultimately dictates the sex of a newly formed embryo. Stripped down and lacking in genetic material, the Y chromosome has the relatively simple job of turning on the sex-specific machinery that is ready and waiting to go in the other chromosomes.

The genetic blueprints for maleness and femaleness—instructions that determine body shape and other sex-specific attributes—are distributed throughout the other chromosomes, from which they are available to either males or females, depending simply on whether or not they receive an activation signal. If sex differences in our species were entirely encoded in the sex chromosomes alone, they would be slim indeed.

Although sex chromosomes are ultimately responsible for maleness and femaleness, at the immediate level hormones are the enforcers of gender. For example, it has long been known that a cow elk given male sex hormone develops antlers; a young hen given male hormones grows a roosterlike comb and wattles, and commences to cock-a-doodle-do. From studies of this type, we know that individuals are surprisingly androgynous, fully equipped with the genetic material to be either male or female. All they need to become normal males and females is a hormonal shove in one direction or the other.

For the first few weeks after conception, human embryos are indistinguishable as to sex. Then, through a complex series of foldings and distinctive growth patterns, fetal tissue differentiates, forming genitalia as well as internal reproductive organs. During the eighth week of development, if the embryo is male, a single gene on its Y chromosome tells the testes to develop, which produce male sex hormone. If the embryo is female, ovaries develop instead. The rest of a human being's sexual anatomy follows from this basic distinction.

It seems that most of the actual genetic difference between men and women—about 3 percent—is concerned with the type and timing of hormone secretions. But these hormones are crucially important in determining our bodies and our brains. The key players are, for men, testosterone and its variants (referred to as androgens), and for women, estrogen and its close relatives, along with a number of other internal chemicals—notably progesterone, prolactin, and oxytocin—which are the primary regulators of ovulation, pregnancy, and lactation. Testosterone is produced largely by the testes, although small amounts are also secreted by the adrenal glands (as a result, women have small amounts of testosterone circulating in their blood). Female hormones are produced by the ovaries as well as by a part of the brain known as the hypothalamus.

As the Brain Goes, So Goes Behavior

Scientists have known for years that hormones determine whether an embryo will have the genitals of a male or a female: if an embryo (regardless of whether it is XX or XY) is exposed to testosterone early in its development, it becomes anatomically male. No equivalent embryonic role exists for estrogen: an embryo does not become female in the presence of estrogen but rather in the absence of testosterone. In other words, femaleness represents the "default" setting in embryonic development; left alone, a human fetus becomes female. Recently, neurobiologists have just begun to understand that these same hormones also influence the growth and organization of an embryo's brain. As the body goes, so goes the brain. And as the brain goes, so goes behavior.

There is nothing very surprising about this connection between body and brain. It would be far more peculiar if evolution had produced bodies and brains that were not in tune with each other. Why have a body for making babies without a brain that predisposes us to have sex and to care for babies once they are born? Why have a body for competing with other males or impressing females without an inclination to do so? Obviously, individuals mismatched in such ways would have low biological fitness and thus would contribute little to future generations.

It turns out that the determining factor in all this is not the embryo's genetic makeup (its sex chromosomes as such) but the hormones to which it is exposed. And so male hormones, especially testosterone, not only oversee the emergence of male genitalia but also prod the embryo's brain to become organized, eventually, as male. Absent such male hormones, an embryo will develop a female brain and, with it, female behavior.

The process of differentiation is marked by two crucial periods for secretion of sex hormones: the first phase, when the embryo is first bathed in hormones, is followed by a second phase, which occurs at puberty. These chemical tides act on the human brain in a two-step process analogous to first exposing and then developing photographic film. The initial in utero exposure establishes the basic pattern, exposing the film, as it were. The second hormonal flood, at puberty, develops the film by bathing it in appropriate chemicals, bringing out or activating the pattern laid down years earlier.

The Power of Male Hormones

There is no getting around it: hormones are crucial in determining sex. The laboratory rat offers interesting insights into the process of sexual development in humans, except that a rat's brain is sexually undifferentiated at birth, about like the human brain at eight weeks after fertilization. After birth, the rat's brain quickly becomes defined as either male or female, depending on what hormones are released. Male hormones produce a male brain; if only a very small quantity of male hormone is present, such as that secreted by the adrenal glands, the resulting brain is female. Therefore if a newborn male pup is castrated and thus deprived of most male hormones, it grows up looking like a male but acting like a female. The animal displays much less aggression than would a normal male rat, and, if injected with female hormones as an adult, will solicit copulation from males, using typical female postures.

The older a pup is when castrated, the less female it will be because exposure to male hormones will have made its brain progressively more masculine.

The extreme example of prenatal masculinization is seen in the spotted hyena. Sexually, this species is the most peculiar mammal on earth. Although spotted hyenas come in two sexes, as do all other mammals, to the casual—and even the not-so-casual—observer, there appear to be only males. Female hyenas display what seems to be a fully developed penis and even a well-filled scrotum. But these genitals are illusory. The clitoris is enormously enlarged, and indeed it even becomes erect during social excitement, looking for all the world like a full-sized penis, and the labia are fused and filled with fatty tissue, making them resemble testicles.

What accounts for such an odd state of affairs? It turns out that pregnant spotted hyenas produce exceptionally large amounts of sex hormone that are readily converted into testosterone by the placenta. Thus, female hyena embryos get a dose of male sex hormone comparable to that received by male embryos. The result? Females that look like males and act like them, too. Female hyenas are ferociously aggressive, and this aggressiveness begins when the animals are very young—in fact, immediately after birth. Newborn hyenas of both sexes are vicious, snarling, murderous little brutes, ready to tear each other apart. Mortality among newborn hyenas is exceptionally high; these animals are in many ways caricatures of all that can be unpleasant about being male. But, in this case, normal females act just as malelike as normal males.

Another indication that male hormones determine both anatomical and behavioral sex differences comes from experiments in which testerone is administered to pregnant rhesus monkeys. Although male offspring thus exposed are normal, females are born with a small penis and scrotum and no vaginal opening. In addition, their behavior more resembles that of young males than that of young females. Human twins in which one member is a boy and the other a girl provide further evidence that testosterone helps masculinize the developing human brain. When both a male and a female fetus are present in a uterus, hormones from the male exert a masculinizing effect on the female "next door," rendering her somewhat malelike in mental processes.

The conclusion is unavoidable: male sex hormone is a mind-altering drug. Perhaps the relatively large role played by testosterone explains why paraphilias are overwhelmingly more common in men than in women. As mentioned in chapter 3, men have less to lose by playing fast and loose, and may therefore be more disposed to make sexual mistakes. And lurking behind that looseness may well be the developmental trajectory of male embryos, which require exposure to male hormones. Females might be less vulnerable to pathological disorders because the route to femaleness involves fewer potential missteps.

When the Dose Is Wrong

No one can ethically experiment on human beings to see what happens when an embryo is exposed to different hormones. But, as it happens, there have been enough errors—both natural and as a result of ill-informed medical practices—to provide a fascinating look at how hormones produce human sex differences.

One interesting defect is known as congenital adrenal hyperplasia (CAH), previously called "adrenogenital syndrome" or sometimes, "androgenital syndrome." It affects about 1 child in 15,000. The adrenal glands of people who suffer from CAH produce unusually large quantities of androgens, a process that begins during fetal development. The results are not especially dramatic if the embryo is male; after all, male embryos are normally exposed to large amounts of androgens secreted by their developing testes. But when female embryos are thus exposed, the consequences are notable.

In a moderate case of CAH, a baby girl is born with external genitals that are recognizably male, although incompletely developed. Internally, however, she has normal ovaries, fallopian tubes, vagina, and uterus and is still genetically XX; that is, female. Her condition can be corrected by surgical removal of her male parts, in which case she grows up to be a woman, fully capable of having children. However—and here is the important point for our purposes—girls and women with CAH are behaviorally distinct from typical females. According to their own reports and those of their parents, they are likely to be tomboys, playing roughly and more competitively than other girls and displaying more interest in sports than in dolls. As adults, they tend to be comparatively unromantic and unusually interested in mechanical things.

A study of seventeen girls with CAH found that they preferred boys as playmates, reported few daydreams about marriage and motherhood, and displayed little concern for clothing or makeup. In short, as a result of being exposed to unusual amounts of male hormone while their brains were developing, they had been pushed in a male direction. Removal of their male genitals had little effect: their brains had already been masculinized.

Describing their observations of children with CAH, Drs. John Money and Anke Ehrhardt of Johns Hopkins University noted:

> All control [normal] girls were sure that they wanted to have pregnancies and be the mothers of little babies when they grew up, whereas one third of the fetally androgenized girls with the adrenogenital syndrome [CAH] said they would prefer not to have children. The remainder . . . did not reject the idea of having children, but they were rather perfunctory and matter-of-fact in their anticipation of motherhood, and lacking the enthusiasm of the control girls.

These observations were made more than twenty years ago. In today's climate of greater gender equality, the girls in the control group might choose differently. But when psychologists Sheri Berenbaum and Melissa Hines of Southern Illinois University repeated the study on young girls in the early 1990s, they found that some things don't change: girls with CAH were more likely to play with trucks, building blocks, and the like, whereas girls in the control group preferred dolls, kitchen toys, and so forth.

Some might think children with CAH exhibit malelike traits because their parents treat them more like boys. But we suspect the opposite: that parents of such children probably emphasize their femaleness in an attempt to make them as "normal" and gender appropriate as possible. Why else would they have had their malelike genitals surgically removed, if not to have the children's behavior correspond to their assigned sex?

In the foregoing cases, CAH was mild and the babies were identified as abnormal only by their external genitalia, which were clearly incomplete. When CAH is severe, however, enough testosterone is produced by the embryo's overactive adrenal glands to create what appear to be normal male genitals. "It's a boy," say the nurses and the obstetrician.

Indeed, nothing seems amiss, and the child is reared as a normal boy. Only later, at puberty, are the concerned parents likely to bring their child to a doctor, worried because their "boy" is not developing into a man: he shows relatively little muscular development, there is little or no sign of a beard or other body hair, and his voice remains high-pitched. Analysis of the "boy's" chromosomes reveals that "his" geno-type is female—a normal XX. Yet neither is "he" fully a female.

Apparently, an extra heavy dose of male hormone can not only pro-duce normal-appearing male genitals but also suppress the develop-ment of ovaries. In such cases, parents and child nearly always opt for maleness; after all, the child has been reared as a male and identifies himself as male. Medical treatment calls for additional doses of male hormone, which the nonfunctional testes cannot provide on their own. The child then develops into a man, despite an XX genetic make-up, but does not produce sperm and thus cannot father children. Clearly, the more male hormones are present, the more masculinizing the ef-fect, with patients who have severe CAH ending up behaviorally male and those with mild CAH being female with tomboy tendencies. In short, exposure to male hormone during development overrides the presence of two X chromosomes, producing a distinctly masculinized individual.

Of course, various degrees of tomboyishness are perfectly normal for girls, along with different levels of "sissyness" among boys. There is a great deal of perfectly appropriate, healthy variation in the behavior of boys and girls, so parents need not leap to the conclusion that their tree-climbing daughters or cookie-baking sons are sexually abnormal; the overwhelming majority are just fine.

When Brains and Bodies Don't Match

During the 1950s and 1960s, physicians noted that some women who had difficulty carrying their babies to term had low levels of the female hormone progesterone. By administering diethylstilbestrol (DES), a synthetic form of progesterone, they significantly increased these wo-men's chances for a successful pregnancy.

Subsequent studies, however, revealed a high incidence of abnor-malities among the children of women who had taken DES during pregnancy. Because DES mimics the effects of male hormones, the girls tended to be pseudohermaphrodites (having both male and female anatomical traits), whereas the boys tended to be effeminate. Although the girls' male genitalia could be surgically removed, like girls with CAH, they were likely to be tomboys. They were also likely to be infertile and to suffer more frequently from vaginal cancer. Although boys whose mothers had taken DES were physically indistinguishable from other boys, they avoided the rough-and-tumble of other boys, were generally less assertive, and were often derided as sissies. Why were these boys so effeminate? It turns out that the DES in their systems inhibited the normal interplay between testosterone and the central nervous system, so they developed brains inclined toward the female pattern.

A few other extraordinary syndromes have been identified. We discuss them not as a medical side show, but because these exceptions help prove the rule that early hormone exposure organizes the developing human brain in either a male or a female direction.

One of the more fascinating of these conditions is known as androgen insensitivity syndrome (AIS). The disorder arises when male embryos have a biochemical anomaly that renders them insensitive to male sex hormones. Although the testes secrete normal amounts of androgens, the child's body does not respond to them. As a consequence, the child, who is genetically male (XY), develops outwardly into a female, with testes that remain inside the abdominal cavity.

At birth, infants with AIS look like normal girls; as adults, they are infertile because they lack ovaries, but they otherwise appear normal, though they tend to be several inches taller than the average woman. If anything, however, women with AIS tend to be more feminine than most: they frequently lack armpit and pubic hair, for example, and are believed to be over-represented among fashion models. (It is rumored that at least two famous female movie stars are genetically XY, but publicity agents—for obvious reasons—are not inclined to acknowledge such matters.)

In his excellent book *Eve's Rib*, which offers fascinating details about how hormones affect sexing of the human brain, journalist Robert Pool tells the true story of Maria Patino, champion hurdler on the Spanish track and field team. It seems that in 1985, Patino, to her consternation, failed a medical examination to confirm her sex. "She," it turned out, was a "he," unbeknownst to all—including Maria herself— until "her" chromosomes were identified under a microscope. Maria Patino has the genetic makeup of a male but because her body is insensitive to male sex hormones, she has the appearance, and the behavioral inclinations, of a female.

In fact, individuals with AIS are so feminine in appearance and behavior that many opt to have their abdominal testes removed and a vagina constructed. "With respect to marriage and maternalism," report John Money and Anke Ehrhardt,

> girls and women with androgen-insensitivity syndrome showed a high incidence of preference for being a wife with no outside job (80%); of enjoying homecraft (70%); of having dreams and fantasies of raising a family (100%); of having played primarily with dolls and other girls' toys (80%); of having a positive and genuine interest in infant care; . . . and of high or average affectionateness, self-rated (80%). Two of the married women each had adopted two children, and they proved to be good mothers with a good sense of motherhood.

Women with AIS are the logical and biological inverse of women with CAH: the former are exposed to virtually no testosterone (actually, they encounter it, but their bodies refuse to notice), whereas the latter get an overdose.

Another remarkable condition, even rarer than AIS, is 5-alpha-reductase deficiency syndrome (5ARDS). Because of a biochemical peculiarity, children with 5ARDS are born genetically male but outwardly appear entirely female, much like children with AIS. They differ from the latter, however, in that their bodies lack the biochemical machinery to convert testosterone into the necessary by-product dihydrotestosterone. Although babies with 5ARDS, like those with AIS, have internal testes, they are outwardly so clearly female that no one suspects they are not.

But puberty marks the beginning of a gender nightmare for them. In a relatively short time, their testes descend and their bodies, now under the influence of large amounts of male hormones, become undeniably male. Their behavior, too, is suddenly transformed and these "girls" become young men almost overnight.

In 1980, the fascinating memoirs of one Herculine Barbin were published. Born in France in 1838 and reared in a convent as a girl, Barbin describes becoming unusually hairy at puberty, noting at the same time that her clitoris had become remarkably enlarged. She also fell in love with another girl. Finally, it was recognized that "she" was now a "he," who was renamed Hercule Barbin and sent to Paris as a railway worker at age twenty-two. Eight years later, Barbin committed suicide. He was almost certainly a victim of 5ARDS.

Even though the genitals of children with 5ARDS do not respond to male hormones during fetal development, enough of a male pattern is laid down within the brain during gestation that when hormones increase during puberty, normal male behavior ensues. In such cases, biology obviously wins out over culture: all it takes is the release of additional male hormones at puberty, acting on a brain that had been organized in utero to be male, and more than a decade of socialization is thrown out the window as the "little girl" quickly becomes a young man.

A related case, one initially touted as evidence that socialization transcends biology, suggests exactly the opposite, namely, the power of embryonic hormones. During the 1960s, an infant boy—one of two identical twins—had his penis accidentally amputated during surgery to repair a fused foreskin. Deciding that it would be too traumatic for him to go through life as a penisless boy, his parents decided to rear him as a girl. They agreed to allow physicians to surgically remove his testes, create a vagina, and provide him with female hormones. By the early 1970s, when the child was preadolescent, his transformation to a girl was hailed a success.

The Joan/John case, as it was popularly known, became renowned as evidence for the plasticity of human sexuality and gender, since the child, after all, was a genetic male. But by the 1980s, a different picture came to light. Joan wasn't so girlish after all. She would tear off her dresses, reject the overtures of other girls, and even try to urinate while standing up—behaviors that emerged before the child was told of her genetic history. At age eighteen, Joan was told of her history. She recounts that she wasn't appalled but relieved. "For the first time everything made sense and I understood who and what I was." She opted for male hormone shots, had a mastectomy, and underwent plastic surgery to rebuild male genitals. John is now married and has adopted his wife's children. Although he is sterile because his testes were removed long ago, his reconstructed penis enables him to have sexual intercourse. The point is that even after being treated as a girl and self-identifying, at least for a while, as a girl, John still retained malelike traits, almost certainly because of his exposure to male hormones before his penis was amputated. As Dr. Milton Diamond of the University of Hawaii at Manoa says in reflecting on the case, "It is the head that holds the primary sexual organ, the source of one's identity, and the organ does not lie."

When Chromosomes Get Mixed Up

Sometimes when a person's sexual identity is uncertain, the sex chromosomes themselves are the culprits, either because there are too many of them or because there are too few. For example, occasionally, a female embryo will inherit only one X chromosome, resulting in a disorder known as Turner's syndrome (genetically expressed as XO). Women with Turner's syndrome outwardly appear very feminine, but lack ovaries and other female reproductive organs. Without ovaries, which normally secrete a small amount of male sex hormones as well as more copious amounts of female hormones, Turner's babies experience what appears to be hyper-feminization of their brains. Indeed, girls with this syndrome are often obsessive about dolls and with frilly clothes and are especially repelled by aggressive or vigorous play. They are also likely to yearn for marriage and motherhood.

Another chromosomal error produces so-called supermales, individuals who inherit an extra Y chromosome and thus have the genetic make-up XYY. As men, they tend to be tall, averaging about six feet in height, and are likely to have low IQs and poor impulse control. They are over-represented in prisons.

In another genetic twist known as Klinefelter's syndrome, females (XX) inherit a Y chromosome. Individuals with this syndrome—genetically XXY—look male, and are taller than the average for men. Yet they tend to develop breasts at puberty, and—not surprisingly—have problems with their sexual identity. They, too, are over-represented in prisons as well as in mental hospitals and among transsexuals, transvestites, bisexuals, and homosexuals. Medical treatment for children with Klinefelter's syndrome includes administration of testosterone, but the timing and dosage are pivotal. If not enough testosterone is given, the child becomes irritable; if too much is given, the child becomes aggressive.

With so many twists and turns of genes, hormones, and behavior, a growing number of social scientists agree that the role of biology in determining sex differences cannot be denied or ignored. The fact that it is now being acknowledged by many whose outlook had previously been nonbiological or even antibiological, speaks volumes. Here is a selection from the third edition of a popular and successful sociology textbook, by J. R. Urdry:

> The first edition of this text, based on information available in 1965, presented a thoroughly sociological explanation of the origin of sex differences in behavior. At that time I argued that sex differences were probably completely determined by socialization, and that any innate predisposition to different behavior by the two sexes was trivial. The information available today invalidates my previous explanations. Evidence on the role of sex hormones in differentiating the behavior of other animals has been accumulating for two decades . . . *It is no longer tenable to believe that males and females are born into the world with the same behavioral predispositions.* [Our italics.]

When Men and Women Think Differently

Men and women may behave differently, but do they actually think differently? There is a growing consensus that they do, at least in subtle ways. Here is a recantation by psychologist Diane Halpern, who once disparaged the notion of intellectual differences between men and women. In the preface to her book *Sex Differences in Cognitive Abilities*, Halpern describes her experiences in teaching classes on the psychology of women:

> At the time it seemed to me clear that between-sex differences in thinking abilities were due to socialization practices, artifacts and mistakes in the research, and bias and prejudice. After reviewing a pile of journal articles that stood several feet high and numerous books and book chapters that dwarfed the stack of journal articles, I changed my mind . . . The data collected within the last few years provide a convincing case for the importance of biological variables.

Even Alfred Binet, French inventor of intelligence testing, had to grapple with the different cognitive abilities of boys and girls when he found that boys were scoring lower than girls on his IQ test. But rather than accept the results at face value, he simply changed the test, re-moving some of the questions at which girls scored, on average, higher than boys and adding a few that boys found easier. His assumption was that boys and girls must be the same in overall intellectual functioning, so that in order to make an accurate test, it was necessary to design one in which boys and girls were, in fact, equal!

Even so, like the proverbial cat that is thrown out the door but keeps climbing back in the window, sex differences keep reappearing. Males consistently do better than females on tests of spatial and mathematical ability, and in map-reading. Girls are superior in verbal ability and re-sponsiveness to stimuli, especially sounds. Of course, these are gener-alizations; some girls are better than most boys at math, and some boys are better than most girls at verbal skills. But the fact remains that in general, boys and girls display different aptitudes.

Nonetheless, scientists have been reluctant to inquire into male–fe-male differences, especially when it comes to cognitive function or IQ, possibly for fear of being labeled sexist. Whereas few would contest that men have greater upper-body strength than women or that women lactate, many would protest efforts to identify differences in intellectual ability. Nonetheless, a few courageous scientists have recently bucked this trend and begun to look at male–female differences in cognition, often with special reference to differences in brain function.*

Test Scores

In a groundbreaking study, psychologists Camilla Benbow and Julian Stanley of Johns Hopkins University examined the performance of 10,000 children in the Baltimore area on the mathematics part of the SAT exam. The results: boys do better than girls—the higher the scores, the greater the gap. These findings generated a firestorm of protest, including criticism that the sample size was too small to be sta-tistically valid. Benbow and Stanley accordingly expanded their sample to 40,000, and got the same results. Twice as many boys as girls scored

---

*Interestingly, a significant percentage of these researchers are women: Laura Allen, Camilla Benbow, Sheri Berenbaum, Melissa Hines, Doreen Kimura, Janet Lever, Diana McGuiness, Christina Williams, Sandra Witelson, to name a few.

above 500; four times as many boys as girls scored above 600; and above 700, there were, on average, thirteen boys for every girl.

Some critics claim these findings reflect the fact that boys take more mathematics classes than girls. But the differences appeared as early as the seventh grade, when boys and girls are still taking the same courses.

Of course, social influences cannot be denied. Recall, for example, the infamous Teen Talk Barbie, launched in 1994, which uttered such memorable sound bites as "Math class is tough." Even though Mattel, Inc., quickly withdrew this offensive phrase from Barbie's prerecorded repertoire, the message Barbie was sending reinforced a classic stereotype. If girls are told that math class is tough—and by no less an authority than Barbie herself—some may take it seriously, so that it becomes a self-fulfilling prophecy. Barbie's predisposition against mathematics is, of course, symptomatic of society's deeper and more widespread prejudices about what boys and girls should be like. No toy company, for example, will ever manufacture a G.I. Joe programmed to announce, "I don't want to hurt your feelings," or a Barbie who growls, "Take that, you scum."

It turns out that average mathematics scores of boys and girls are not all that different, because even though boys are substantially more likely to be at the upper end of the curve, they also hold down the lower extremes. In other words, when it comes to math, boys are more likely to be dunces as well as geniuses; girls are more likely to be, well, average. If the number of high-scoring boys reflects social pressures encouraging them to be good at math, then why should there be more boys at the lower end as well? It isn't just the graduate seminars in advanced calculus that are full of boys; so are remedial classes. We don't have an answer to this puzzle, except to note that it may reflect the general tendency of males to be more risk-taking and extreme.

The World of Achievers

Another question arises concerning mental differences between men and women. Why, if women are superior to men in verbal ability and responsiveness to stimuli, are they so poorly represented in the historical panorama of intellectual achievement? The answer that immediately comes to mind is that women simply haven't had the opportunities men have had, for a number of reasons, including male dominance and suppression of women's talent, the rarity of role models, and the primal demands of motherhood.

Author Tillie Olsen—herself a Depression-era high school dropout whose writing career was sacrificed while she raised four children and worked at various menial jobs—has argued that motherhood makes sustained intellectual creativity exceedingly difficult. She notes that prior to the twentieth century, there are few examples of creative women who were not single, lesbian, cloistered, wealthy, or, at the very least, childless.

Mothers who pursue an intellectual passion too intently have often been criticized for not being sufficiently devoted to their children. Richard Wagner could spend thirty-four years completing his Ring Cycle but if Cosima Wagner, herself artistically inclined, had devoted equal time to her own creative intellectual project, she would have been an oddity, almost certainly derided as having abandoned her station in life. When Herr Wagner did just this, he was applauded as being focused and hardworking. Frau Wagner's "ring cycle" was her wedding band.

In virtually every field, the argument goes, there have been numerous great masters but precious few great mistresses. No female artist compares with da Vinci, Rembrandt, Michelangelo, Goya, Matisse, or Picasso. No female composer occupies the same rank as Vivaldi, Mozart, Bach, Beethoven, Wagner, Tchaikovsky, or Rachmaninoff. Where are the female writers who rival Shakespeare, Molière, Tolstoy, or Goethe? Or female scientists who stand beside Newton, Copernicus, Harvey, Kelvin, Darwin, Einstein, or Pasteur?

Of all the arguments for male–female differences, this one strikes us as particularly weak. Until social systems grant equal access to women and provide equal encouragement of their talents and inclinations, we cannot conclude that women would not be as successful in the upper ranks of creativity as men. It is said, for example, that Mozart's sister, who died very young, showed as much talent as her brother. We suspect that for every identified male genius, there are many females whose talents have been redirected or squelched by society. Marie Curie, for instance, despite winning two Nobel prizes, was never admitted into the Paris Academy of Sciences (the first woman didn't get in until 1980). And it wasn't until 1997, faced with international outrage and the threat of boycotts, that the Vienna Philharmonic Orchestra agreed to hire its first female musician, having previously claimed that admitting women to its august ranks would alter its unique sound! There may be some truth to the adage that women have to be twice as good as men to obtain comparable recognition. Indeed, the fact that even small numbers of women have achieved renown in traditionally male-dominated fields—writers such as Jane Austen, Virginia Woolf, and Emily Dickinson; anthropologists such as Margaret Mead and Ruth Benedict; politicians such as Margaret Thatcher; artists like Georgia O'Keeffe—might suggest a certain superiority.

Honest open-mindedness nonetheless leads us to ask whether there might be genuine male–female differences in thinking, ones that manifest themselves in different male–female accomplishments. Might there be, for example, a spatial component to the theory of relativity, to Copernicus's and Ptolemy's conceptions of the solar system, and even to the structure of music composition, that gives males a slight advantage? In addition, might not male aggressiveness, which stems from male strategies for sexual success, help propel men to greater prominence? Most accomplishments require a kind of pushy, persistent determination in addition to raw ability. Nice guys, we are told, finish last.

Another factor is men's desire to impress women, an offshoot of sexual selection in which males must compete among themselves for access to choosy females. Is it architectural genius that induces the male bowerbird to create his magnificent structure during courtship? Is it musical genius that contributes to a warbler's song or an elk's bugling? Certainly, the male bowerbird relies more on genetic instructions to construct his bower than Frank Lloyd Wright ever did to design the Guggenheim Museum, and the male warbler shows far less creativity and originality than did Beethoven. Nonetheless, there may be an unrecognized sexual component—in short, a penchant for competition and showing off that is particularly male—underlying many of the accomplishments for which some men have achieved so much renown.

The irritating matter of "woman's intuition" carries with it the condescending implication that women, being somehow more intuitive, are also less cognitive. But there may be something to the concept after all. Sociologists have long known that subordinate individuals tend to be highly sensitive to the nuances in behavior and mood of their social superiors. Such sensitivity makes evolutionary as well as social sense because the success—even, on occasion, the survival—of subordinates may depend on their reading the dominant individual's moods and inclinations correctly. Like blind people who develop an acute ability to interpret sounds, social subordinates may develop an acute ability to interpret the intentions of others, especially those of their social superiors. By contrast, dominant individuals can be relatively oblivious to what is going on beneath the surface. Office employees may fret over a casual remark made by the boss or worry about whether he or she is in a good mood, but the boss rarely thinks much about the moods and nonverbal attitudes of those down the ladder.

To some degree, women's sensitivity to others may also stem from their role as caregivers for infants, who must make themselves understood without words. Women who respond quickly to the needs and desires of their children almost certainly increase the likelihood that those children will survive. We suspect that if roles were reversed so that men did most of the child care, they would be the ones with intuition. But our suspicions can never be tested, since there is no society in which the child-care roles of men and women are reversed.

Philosopher Jean-Paul Sartre suggested that there is a sexual component to how we go about understanding the world. "Knowledge," he wrote, "is at once a penetration and a surface caress." Is that why science, at least so far, has been largely a male province?

As social barriers to women in science continue to fall, the number of female scientists is rising proportionately. It will be interesting to see whether the nature of scientific inquiry changes as a result; that is, will a more womanly style of science emerge, less interested in penetrating and more concerned with the subtle unraveling of relationships?

Suggestions that such a shift may occur come from field studies of nonhuman primates, a research area that has been revolutionized by the work of three pioneering female scientists in particular: Jane Goodall, studying chimpanzees; Birute Galdikas, studying orangutans; and Dian Fossey, studying gorillas. In all three cases, the hallmark of their work was patient, nonintrusive watching and waiting, combined with unflagging attention to the details of the private lives of each of their subjects. In their early stages, the efforts of these scientists were derided—often, by male colleagues—as "soap opera science," too gossipy and personal in focus and insufficiently concerned with accumulating a large, statistically significant sample.

Why Men Read Maps and Women Ask for Directions

The question of a female style in science—more subjective, more attuned to subtle interrelations, more focused on individuals than on abstract principles—is open to debate, but other differences in intellectual style are clear. As already mentioned, men reliably score higher than women in tests involving spatial skills, wheras women beat men when it comes to verbal and auditory capabilities. One clever bit of research highlighted these differences. Men and women were asked to go through the alphabet and identify those capital letters that fell into one of two categories: those having a curve in their shape, like the letter *S* and those having a long *ee* sound, like the letter *T*. (Some, like *B*, have both, and some, like *M*, have neither.) When it came to the shape task (related to spatial visualization), men made fewer mistakes and finished the task more quickly than women; on the verbal task—identifying those with the *ee* sound—women did better.

Their greater aptitude for spatial relationships makes men more adroit at reading maps and perhaps at mathematical functions generally. It has also been found, and repeatedly confirmed, that women rely more on landmarks for finding their way whereas men prefer navigating with the help of maps and compass directions. Additional support for the biological basis of cognitive abilities comes from animals; namely, rats. Male rats apparently form an internal mental map to navigate a maze, whereas female rats orient themselves via conspicuous objects. Moreover, early exposure to testosterone makes female rats more malelike in their navigating strategies, whereas castrated male rats are more femalelike in their homing behavior.

More evidence comes from the influence of sex hormones on subtle cognitive functions such as verbal fluency, verbal memory, and manual dexterity. For example, a woman's menstrual cycle and its associated hormonal fluctuations influence her spatial ability and verbal fluency—though not in an overwhelming way, mind you. The observed cognitive effects, although real, are small and are detectable only by sophisticated psychological testing, in which subjects are asked to rotate images mentally, to recall different kinds of words, and so forth. In short, when female hormones are high, female-biased traits such as verbal fluency, verbal memory, and manual dexterity are also high; however, spatial ability—a trait in which men have a statistical advantage—is particularly poor at these times. As one might expect, testosterone appears to improve those traits, such as spatial ability, that are generally masculine.

The relationship is not simple, however, because estrogen and testosterone are not direct opposites in their effects. Too much testosterone—even in the presence of estrogen—can impair clear thinking, and although there is evidence that estrogen improves memory and acts as an antidepressant, it can also act as a tranquilizer. In addition, although in low doses testosterone is a tonic, or an activator, in high doses it can lead to "androgen psychosis," with symptoms such as anxiety, paranoia, and heightened aggression. Testosterone can also be combined with estrogen to elevate sexual desire in postmenopausal women. Interestingly, some women treated with testosterone for this reason report more energy, easier weight loss, and better spatial perception as well. To make things more complicated, testosterone changes into estrogen as part of normal female metabolism. Figuring out the exact details of how hormones influence the human brain will keep researchers occupied for years to come.

Inside the Brain

Even though the brains of men and women function differently, few anatomical differences have been found, and most of them are hotly disputed. One difference, however, is undeniable: a man's brain is bigger. It measures about 88 cubic inches on average, whereas a woman's measures about 77 cubic inches. At one time, this difference was said to prove the mental superiority of men; however, the fact that larger individuals almost always have larger brains had been conveniently ignored. In fact, when corrections were made for the difference in body size, a woman's brain was revealed to be a bit larger than a man's. But regardless of the sex, the human brain is undoubtedly large. The cranial capacity is variously attributed to tool use, communication, hunting, and so forth, in various combinations. An additional evolutionary rationale for the large human brain might be found in the phenomenon of female choice: if prehistoric women preferred to mate with brainy males, this would have produced more brainy offspring than their less astute cohorts, contributing to the explosive growth in human brain size that characterized the early evolution of *Homo sapiens*.

In any event, the roughly 10 percent smaller size of female brains continues to puzzle neurobiologists. In an absolute sense, shouldn't this mean that women would be 10 percent less intelligent than men? Clearly, they're not. Sandra Witelson of Canada's McMaster University has recently come up with a partial explanation. While conducting postmortem examinations of brain tissue, she and her team found that in a part of the temporal lobe known as the planum temporale—a center for language and auditory function—the neurons of women's brains are more tightly packed than those of men's brains. In other words, women have the same number of neurons (which presumably hold the key to intellectual functioning) as men; they're just more densely organized. It remains to be seen whether sex differences in neuronal density occur in other brain regions and what effects, if any, such differences have on male–female differences in behavior. As Witelson so aptly put it: "The female brain is not just a scaled-down version of the male brain."

Witelson's statement should not be altogether surprising. After all, the female body is not just a scaled-down version of the male body, just as female strategies and tactics for evolutionary success are not just scaled-down (or up!) versions of male strategies and tactics. By the same token, the brains of male and female animals are known to differ when such differences contribute to reproductive success. For example, the part of the brain that controls singing in songbirds is much larger in males than in females; similarly the region of a rat's brain that influences spatial learning is larger in males, which typically wander more and thus have more need to negotiate complex physical terrain. Although fascinating, such findings have not been especially controversial.

But sex differences in human brain anatomy have now begun to appear, and they have been especially troublesome for those concerned that discovery of male–female differences in brain structure might portend dire social or philosophical consequences. In the brains of rats, for example, there is something known as the sexually dimorphic nucleus (SDN). This structure, which lies in the hypothalamus not far from the region influencing aggression, is also thought to influence sexual and maternal behavior. Notably, it is three to seven times larger in the brains of males than in those of females. Not surprisingly, early exposure to sex hormones accounts for the difference: inject testosterone into infant females, and they develop male-sized SDNs. If males are given a testosterone-blocking agent, their SDNs shrink, resembling those of normal females. Several years after the discovery of the SDN in rats, an equivalent structure was found among human beings, and it, too, is bigger in males than in females. But to date, its functional significance is unknown.

Excitement as well as controversy have swirled around a report of brain differences between gay and straight men. In 1991, neurobiologist Simon LeVay reported on a particular region in the hypothalamus known as INAH-3, which was said to be twice as large in men as in women. After examining a number of autopsied brains, LeVay announced—to substantial media attention—that INAH-3 in homosexual men resembled that of heterosexual women, being about one-half the size of the same region in heterosexual men. Many aspects of this finding remain unclear, such as whether it will hold up when a larger sample of brains is examined and if it does, whether this brain difference determines sexual orientation or is simply one of many side effects of male homosexuality.

Political conservatives, who often consider homosexuality a lifestyle choice (and in their judgment, something to be morally condemned), are discomfited by the notion that sexual orientation may have a biological basis. In contrast, liberals, who tend to dislike biological discussions of behavior (because they raise the specter of discrimination, if not eugenics, and because the term biological is often erroneously equated with *unchangeable*) are comforted to think that homosexuals should not be blamed for "doing what comes naturally."

Interesting anatomical differences have also turned up in the brains of transsexuals, individuals who live with the agonizing certainty that something is wrong with their sexual identity. Most transsexuals describe being women trapped in men's bodies; more rarely, men trapped in women's. So intense is their inner sense of gender that many are desperate for radical sex-change surgery so as to match on the outside what they have long felt on the inside.

A group of researchers in the Netherlands reasoned that if transsexuals really do reflect, say, female brains stuck in male bodies, something useful could be gained by comparing the brains of transsexuals with the brains of heterosexual and homosexual men and women. The scientists from the Netherlands Institute of Brain Research in Amsterdam, looked carefully at autopsied brain tissue, paying special attention to parts of the hypothalamus known to coordinate sexual behavior and reproductive hormones.

The results? Dramatic differences in the size of one region, known as the "central subdivision of the bed nucleus of the stria terminalis." Among heterosexual and homosexual men, this region averaged about 2.6 cubic millimeters in size; among women, it average 1.73 cubic millimeters; among transsexuals, the average dropped to 1.3 cubic millimeters. The exact significance of these findings has yet to be determined, but the prominence of this one structural difference suggests that there are others to be discovered.

Using Our Heads

Whatever the structural distinctions between the brains of men and women, it is increasingly clear that men and women *use* their brains differently. The human brain, for example, is not "ambidextrous." Just as most of us are either right- or left-handed, our brains have right and left sides, with different specializations.

But suggestions that males are more "left brained" and females more "right brained," boil down to a good myth. The left side is, in fact, especially concerned with verbal tasks—a female specialty—whereas the right side deals largely with visualizing or manipulating objects in space, a male specialty. Still, the picture is clouded. The left side of the brain also specializes in analytical and linear thought, supposedly something of a male advantage, and the right side deals with nonlinear, holistic thinking, widely considered a female specialty. Still, the portion of the right hemisphere devoted to visual-spatial tasks is larger in men than in women. Possibly, men are more limited in verbal access to their emotions as a result. Put another way, women simply may have more neurons available to connect words with feelings.

Recently, it has been found that brain lateralization is more intense in men, indicating that women are literally more "scatterbrained"—in an entirely different sense from what that term usually implies. Put another way, brain function in women is thought to be less specialized—

less limited to specific brain regions, while the brains of men are more rigidly dedicated to one task or another. The lateralization notion arose partly because men are more likely than women to lose mental function, especially language capacity, as a result of brain injury, a fact that suggests men's brains are more compartmentalized.

A team led by Bennett Shaywitz and Sally Shaywitz of Yale University provided what we believe is the first clear documentation that the brains of men and women are indeed arranged differently. The researchers gave men and women a variety of problems involving language skills such as letter recognition, rhyming, and semantic categorization. Then, using magnetic resonance imaging (MRI), the Shaywitzes compared patterns of cerebral blood flow in the two sexes. They found that in men, activity was lateralized in a particular brain region known as the left inferior frontal gyrus. In women, the activated regions were more diffuse and were located in both the right and left sides of the brain.

There are many ways, supposedly, to skin a cat. We wouldn't know; we've never tried. But it now seems indisputable that when it comes to language, there are at least two different ways for the human brain to arrive at the same result, and in this respect, at least, male brains and female brains part company.

This realm of study remains highly controversial, however, with some brain researchers maintaining that women's brains are, if anything, *more* specialized, especially in regard to language. The fact that women recover the facility of speech more readily than men after strokes and brain injuries may be largely because their language centers tend to be in regions that are somewhat less vulnerable to damage. The brain's response to injury represents a very active area of research, one in which new findings are published almost daily. Perhaps the safest generalization is that differences—of some sort—exist, and will be increasingly revealed as time goes on.

Whatever the precise structural and organizational differences may be between a man's brain and a woman's brain, we would like to suggest an evolutionary rationale for the male advantage in spatial-mathematical skills and the female superiority in verbal/emotional competence. Women, being the choosier sex, are likely to enhance their evolutionary fitness by achieving greater intimacy with one mate, who is likely to stay around, rather than by seeking numerous mates. (When has a woman ever complained that her husband wants too much emotional intimacy, or a man that his wife gives him too much freedom?) Men, however, are likely to increase their reproductive success by keeping some of their emotional energy to themselves. By holding back on intimacy, men may therefore be better able to participate in relationships with more than one woman.

Men can and probably should learn better ways of communicating their emotions (more precisely, with words rather than fists). But insofar as the male brain is less predisposed than the female brain to verbal communication, women should understand that performing emotionally may be difficult for some men. Psychiatrists have a word for the inability to express emotion verbally—alexithymia, meaning literally "without words for feelings." Maybe men "say it with flowers" because they have a hard time saying it in other ways.

Uncertainties

Everyone agrees that there are some clear-cut biological aspects of being a woman: menstruation, gestation, lactation, all narrowly associated with reproduction. It has been argued that these specific reproductive events represent the only true male–female differences and that even these (the pregnancy-related ones) apply only to those women who become mothers. Furthermore, it is said these traits are relevant for only a brief time, so their importance—when applicable at all—is very limited. But such thinking misses a crucial point: biology primes and prepares people for behaving in certain ways whether or not they actually do so, as long as there is a high probability that they might. In the same way, each month, a woman's biology prepares her uterine lining to receive an embryo, whether one appears or not, that month or ever in her life. Similarly, a man's biology makes him somewhat more likely to behave aggressively, with sexual avidity, and so forth, whether or not his circumstances or moral beliefs ever provide occasion for such behavior.

At the same time, of course, there is plenty of room for the environment to affect the body: not only muscles and bones, as we know from the effects of exercise, but even brains. It is well established, for example, that the brains of rats become heavier and more complex when the animals grow up in interesting and diverse environments. The brain is not a muscle, strictly speaking, but mental activity helps keep it strong and effective. In addition, the brain can be strongly influenced by its environment. Men can be encouraged to become more nurturing, less violent, and more emotionally expressive. And women can be encouraged to become more adventurous, more aggressive (as witnessed by the effectiveness of assertiveness training), and better at spatial relations.

The brain is a mechanism, a device for producing behavior, although it remains a great mystery how a set of genes and hormones dictates what the brain actually does. Someday, we expect, geneticists and neurobiologists will finally unravel the question of how genes directly influence behavior. Until then, biology's reach will continue to exceed its grasp—which is, perhaps, the way it ought to be.

CHAPTER 9

The Power to Choose

IN THE END, a woman, as a man, has the power to choose, and to make her own heaven or hell.
— Betty Friedan,
The Feminine Mystique

*I*n making sense of our sexual differences, we are left with the question, So what? The Serenity Prayer, written by theologian Reinhold Niebuhr, comes to mind: "God grant me the serenity to accept the things I cannot change, courage to change the things I can, and wisdom to know the difference." Admittedly, Niebuhr wasn't thinking about sex differences when he wrote the Serenity Prayer. But his sentiments are beautiful and appropriate.

Our goal in writing this book has been to provide information, to dispel confusion, and, where appropriate, to encourage the serenity that comes from understanding. We deeply believe that by understanding the tides of their own biology, from lusting to nesting, from male–male competition to female choice, people will achieve greater self-knowledge and the ability to make better personal and social choices. At the same time, we firmly reject the notion that biology is destiny. We believe that human beings ought to come to terms with their biology as informed participants rather than as helpless victims.

Just Say "Know"

One of the most famous—and least satisfied—injunctions in Western thought is the ancient Greek command "Know thyself." It is the spirit in which we have written this book. Or, as Alexander Pope put it, and to which we unblushingly add a modification, the proper study of mankind is man . . . and woman. Thus, we firmly believe that only by cultivating conscious awareness of the human condition can we become less the servants of our genes and hormones, and more the masters of ourselves, our culture, and our lives.

Earlier we stated that biologically, human beings are rather ordinary mammals. The human claim to specialness rest on our immense brain power and the degree to which humanity's behavioral repertoire includes cognition, culture, symbolism, language, and so forth. As a result, we are almost certainly less constrained by our biology than is any other species.

Moreover, as conscious, thinking creatures, people have the luxury of saying no to many of their biological tendencies. After all, tendencies are just that: they are inclinations or propensities, not commands or mandates. Thus, a naturally shy person can be taught assertiveness, a bully can learn to channel his aggression in positive ways, a philanderer can opt for monogamy, and a monogamist can gain insight into his philandering impulses.

We would even suggest that a fundamental part of being human is the ability to say no to the genetic whisperings within us. Each person is to some degree preprogrammed, but as a reflective, conscious being he or she can pick and choose which programs to use and which to deny. Human beings have even devised ethical and social systems to help decide when such nay-saying is appropriate. In short, we are not tabulae rasae, blank slates upon which parents, teachers, and society inscribe at will, nor are we DNA-driven automatons destined to play out our lives as genetically controlled robots.

Early in the twentieth century, eminent British biologist Julian Huxley urged people to avoid the fallacy of "nothing but-ism," the notion that because human beings are animals, they are nothing but animals. His warning, although well taken, can be turned around with equal cogency. Although people are strongly influenced by social learning, early experiences and cultural traditions, this does not mean that they are "nothing but" the sum total of their social learning, early experiences, and cultural traditions. Complex and flexible as we are, and capable of saying no as we may be, we are still not entirely immune to the influences and tendencies that compose our shared biological heritage.

Vive la Différence

It is said that during a debate over women's rights in the French Parliament, a lawmaker expostulated, "After all, there is a difference between men and women," whereupon the members of the Parliament rose as one and declared, "Vive la différence!" This story is almost certainly apocryphal, but we subscribe to the sentiment. How could anyone advocate multiculturalism and celebrate the differences among ethnic groups while denying the much more important differences between men and women? Similarly, anyone who cares about liberation might want to embrace not only the liberation of women from harmful stereotypes but also the liberation of all people from rigid ideologies. Just as devotees of multiculturalism urge us to celebrate cultural differences, there is good reason to celebrate fundamental differences between the sexes.

The material in this book is controversial not because it is untrue but because it can be misused. We hope that our arguments will not be construed either as ammunition for sexism or as fatalistic capitulation to the "natural order." In describing the biology of maleness and femaleness, we focused on what *is*, not what ought to be. Although we are, in fact, greatly concerned with what ought to be, it is not something we derive from the study of evolution.

More than two hundred years ago, Scottish philosopher David Hume warned about confusing "is" with "ought," an error later identified as the "naturalistic fallacy." Most of us accept that a lion preying on a zebra is neither good nor evil and that there is no moral substance to the colors of a rainbow or the power of a hurricane. These things simply *are*. The simple fact that something is natural does not make it good or bad, even though many thoughtful people agree that there is something laudable about natural foods, natural childbirth, or a natural environment. The sad truth is that natural things can also be pretty nasty, as demonstrated by gangrene, typhoid, and AIDS. In such cases, the artificial is preferable. "Smallpox is natural," wrote Ogden Nash. "Vaccine ain't."

For our purposes, the important point is that evolutionary biology sheds considerable light on male–female differences but says nothing about whether these differences are good or bad and whether they should be exaggerated, celebrated, ignored, or deplored. Meteorologists don't worry that by attributing the formation of hurricanes to such natural phenomena as moisture, heat exchange, and the Earth's rotation, they are promoting the destruction hurricanes often wreak. So it is—or should be—with those who seek a natural understanding of sex differences.

In seeking the truth behind the gender gap, skeptics need to outgrow "Brahean thinking," a term we take from Tycho Brahe, a renowned astronomer of the sixteenth century who came up with a model for the solar system that was incorrect but revealed much about the human penchant for wishful thinking. Brahe could not deny the evidence that the five planets then known (excluding Earth)—Mercury, Venus, Mars, Jupiter, and Saturn—circled the sun. At the same time, he could not accept that Earth also revolved around the sun instead of enjoying God-given centrality. So Brahe came up with a compromise: he devised an astronomic model whereby the five known planets obediently circled the sun, as the scientific evidence demanded, but then proposed that the sun and those five planets circled Earth!

Brahean thinking thus involves a gradual, grudging acceptance of facts—what one *knows* to be true—while stubbornly retaining what one *wants* to be true. Thus, many intelligent, well-informed persons acknowledge the legitimacy of evolutionary biology and even its relevance to human beings on the one hand, but on the other, they adopt worldviews that deny that evolution matters. More specifically, they acknowledge the existence of basic (that is, anatomical and physiological) male–female differences yet stubbornly insist that these differences don't matter!

On the other side are those who attribute every aspect of sex differences to the inexorable unfolding of human biology, including an array of phony gender gaps that almost certainly derive more from ideology than biology. We say to extremists on both sides, whoa, there; ease up. Both learning and biology make us what we are today. Genes and experience interact, and only together do they shape biological structures and behaviors. Knowledge of both—genes as well as experience—is necessary to understand ourselves and to make good choices in our personal lives and in society.

Difference Feminism

We are aware that in discussing male–female differences, we run a risk: not so much the straightforward risk of incurring the ire of those who cannot abide the notion of *any* distinction between the sexes but rather the more subtle risk that we might unintentionally be contributing to society's penchant for genderosclerosis, hardening of the categories.

Why, then, did we choose to handle such a hot potato? In part, it is because of our fondness for what we take to be the truth. But we undertook this work for many other reasons, too: Because we are confident that evolutionary biology holds the key to many vexing questions about human nature and our behavior. Because we cannot relinquish the search for such knowledge to those less respectful of its limitations. Because we are convinced that men and women don't have to be seen as identical in order to deserve equal respect and equal opportunities. Because male–female differences are interesting and important and because—like mountains to be climbed—they are there. Also, in part, because we hope that a better understanding of our differences will help generate a richer and more nuanced grasp of ourselves and of what we all share, of our needs and aspirations and the social rules and restrictions against which we struggle, men and women alike. In short, we aim for a deeper grasp of our common humanity.

We worry less that we might be misjudged by our friends than that we might inadvertently provide ammunition for our opponents, that our work might add to the stereotypes on which misogyny and sexism have in part rested. As we have noted, thoughtful feminists have good reason to be wary because phony biology has long been used to buttress the dominance of men: witness the biologist as apologist for social evil or the ideologue masquerading as interpreter of cold reason and scientific fact.

In 1992, for example, during a heated debate over whether female priests should be ordained, Austin Vaughan, auxiliary Catholic bishop of New York, stated, "[To have] a woman priest is as impossible as for me to have a baby." He was confusing biological with social reality. Given the current state of endocrinologic and obstetric science, it is indeed biologically impossible for a man to have a baby. But, *pace* Bishop Vaughan, it is definitely possible for a woman to be a priest. (Episcopalians, doctrinally as well as biologically similar to Roman Catholics, have them.) Male childbearing is currently beyond our reach, but the ordaining of female priests is well within the human evolutionary repertoire. If tomorrow the Roman Catholic Church were to decree, "Let there be female priests," there would be. Sadly, not only is Bishop Vaughan guilty of stating a social preference as though it were biological law, but he also has had lots of company.

Indeed, such misinterpretations and misunderstandings of male–female biology have led sociologists such as Cynthia Fuchs Epstein to warn against "deceptive distinctions." We understand their worry. After all, arguments derived from biology have long been used as clubs with which to beat women into submission. We are feminists ourselves, painfully aware of the history of sexual discrimination and of the way male–female differences, wrongly characterized, have contributed to injustice. Not surprisingly, many feminists have chosen to deny the very existence of sex differences.

We find that those most stridently engaged in this denial—who insist that there are no real differences between men and women—are overwhelmingly people whose learning derives from a narrow range of academic training, chiefly traditional sociological and psychological theory, or whose worldviews reflect a political or social agenda. Those whose learning comes from life rather than from textbooks or ideology are far more apt to take sex differences for granted.

Our concern has been to explain and elaborate on the distinctions we see as real, distinctions that have recently been acknowledged by a number of researchers, including many leading feminists. Educational psychologist Carol Gilligan and linguist Deborah Tannen, for example, have written extensively about male–female differences. Their findings reinforce each other when seen in the clarifying light of evolutionary biology. Yet both of these fine scholars have studiously refrained from examining the reason for those differences. Tannen, for example, in *You Just Don't Understand*, discusses male–female patterns of conversation but never explores how these stylistic differences came to be or what purpose they serve. Still, many Americans have found the mere existence of such distinctions eye-opening, giving legitimacy to Tannen's plea for cross-cultural understanding across the gender gap.

In her book *In a Different Voice*, Carol Gilligan presents a cogent argument for male–female differences in moral development, pointing out the pitfalls of following a male-based model for social behavior and ethical responsibility when women and girls are likely to follow different models. The result? Not a better or a worse voice, just a different one. But once again, although Gilligan explains how male and female moral voices tend to be distinct, she never asks why.

Feminists tend to fall into two groups. One group, which includes Tannen, Gilligan, and others, acknowledges and even revels in male–female differences; the other group, in contrast, believes that any such recognition is dangerous and ill-advised, providing potential aid and comfort to the sexist, patriarchal, woman-oppressing Enemy. We side with the former group, practitioners of what poet and essayist Katha Pollitt has called "difference feminism."

Indeed, difference feminism has been especially embraced by those eager to portray women as *more* caring, nurturing, and relationship oriented—in short, in many ways superior to men. In the process, some difference feminists probably go too far, sentimentalizing the peaceful, life-affirming inclinations of women in general and of mothers in particular. Accurately portrayed, however, difference feminism has great potential. It could lead to a new argument for affirmative action. If it is accepted that there are genuine, valuable differences between men and women, then there is more justification than ever for hearing women's voices in every field of endeavor, whether it be the arts, science, law, politics, economics, or any other. If, however, men and women are taken to be interchangeable, presumably it wouldn't matter if we hear from only one sex . . . and given the greater aggressiveness of men, the likelihood is that this one sex would be male.

Opponents fret that if difference feminists have their way, women might find themselves typecast with a variety of arbitrary, misleading male–female distinctions. Some worry that a "different voice" might seem too weak or high-pitched to be taken seriously; others argue that it is better to speak differently than to keep quiet altogether. Our view is more optimistic: we believe that when sex differences are properly understood, they will be seen as truly complementary, like the yin and yang of Eastern wisdom, with neither one dominant or in any way preferable and with the unified whole far more interesting, exciting, and complete than either part taken alone.

Clear-Eyed Understanding

"Why can't a woman," asked an exasperated Henry Higgins in *My Fair Lady*, "be more like a man?" The answer is simple: a woman *can* be more like a man. And a man can be more like a woman. But a gap will always remain. In fact, men and women are more different from each other than are males and females of any other species. Not biologically—after all, there are certain deep-sea fishes in which the male is a tiny parasite that spends most of his life within the female's genital opening—but culturally, in the diversity of human institutions and ways of life.

Winston Churchill once described a political opponent, Clement Atlee, as "a modest man with much to be modest about." We in turn note that men and women are so extraordinarily different because human beings have so much to be different about.

How many different ways are there of being human? If we could live our lives over and over again, how many different lifestyles might we adopt during our many lifetimes? A hundred? A thousand? Very many, to be sure, although the possibilities are not infinite. The great variety of human cultural traditions is impressive, so much so that cultural anthropologists have been kept busy for decades merely cataloging the variations.

But our point is that such variations follow a limited number of themes. Just as a family selecting a Christmas tree will focus on the differences among trees (this one is taller; that one has more branches or is shaped more symmetrically), observers of human behavior tend to focus on the differences between, say, the lifestyle of a New Guinea highlander and that of a New York taxi driver. Yet imagine an objective Martian scientist considering a Christmas tree farm for the first time; this individual, almost certainly, would immediately be struck not by the differences among the trees but by their similarity. By the same token, an objective, biologically based view of human social arrangements cannot help but notice the fundamental *similarities* that underlie the superficial cultural variations.

Among these cross-cultural similarities, our Martian scientist would most likely be impressed with the persistence—almost the monotony—with which the same basic male–female patterns are expressed over and over again. He or she (or it?) would vainly search for societies in which women were consistently more violent than men, or men took primary responsibility for child care, or women rather than men competed for social and political status, or men were coy and shy while women were pushy sexual adventurers. For all our inventiveness and imagination, we as a species have chosen to express our maleness and femaleness within a narrowly circumscribed range of options.

Some would say, "If men are more likely to be violent or if women are more likely to do the parenting, it is because that is how society has been organized, not because of the nature of men and women." But such statements ignore the bigger questions: *Why* are societies organized in this way? And why are they *all* organized in the same basic way? Also, is it a coincidence that similar patterns are found among other, similar animals and that those differences are consistent with biological theory?

Biologist E. O. Wilson suggests that "genes hold culture on a leash." In other words, biology sets certain limits, although within those limits, social norms hold sway. No society expects men to give birth or even to take the primary role in caring for young children, and no society expects women to make up the bulk of its military combat force. We predict, moreover, that no society ever will.

Those who oppose the "biologizing" of sex differences point to the danger of reductionism. What frightens them so much? Some would say that comparing the intricacies of human behavior to the antics of honeybees, prairie chickens, and baboons is somehow belittling.

Then there is the supposed problem of genetic determinism, the danger that attributing aspects of human behavior to genes might deprive humanity of free will and diminish its prospects for change and betterment. But genetic *influence* is a far cry from genetic *determinism*.

We do not claim that males and females differ in all ways or that such differences, when they occur, are necessarily rooted in evolution. We do maintain, however, that sex differences are real and widespread and that an evolutionary perspective provides crucial intellectual leverage to enhance an understanding of human nature and relationships.

Let us be clear: it does not demean women to assert that they are different from men, any more than it diminishes men; it simply helps us make sense of the two sexes. Moreover, the argument that sex differences necessarily condemn women to second-class status contains the hidden presumption that males are the defining sex—that if females are not identical to males, they are by definition inferior. There is no rational basis for a belief that men, rather than women, represent the essential archetype for humanity. We are in fact a species composed of two sexes, different but inextricably linked and essential to each other.

For example, making sense of sex differences leads to the conclusion that men are likely to employ an aggressive style, an evolutionary, prehistoric payoff for those successful in male–male competition. Possessing this style, and to some degree being possessed by it, men, it seems, are pushy creatures whose aggressive tendencies steamroll whatever lies in their path: other men, women, puppy dogs, rain forests, and so on. Thus, we submit that most of the time men are concerned less with oppressing women as such (that is, engaging in misogyny) than with oppressing, suppressing, or repressing *anyone* who gets in their way.

Still, no one venturing into the realm of male–female differences should ignore the reality of gender-based inequality. For example, the World Health Organization reports that women constitute slightly more than one-half of the worldwide human population, make up one-third of the paid labor force (the labor of child care and homemaking being almost universally unpaid and undervalued), and are responsible for two-thirds of the hours worked. Yet women receive only one-tenth of the world's annual income and own less than one-hundredth of the world's property. Something is dreadfully wrong.

That "something" resides less in the biological reality of sex differences than in the way human beings have chosen to organize their societies. But maybe a clear-eyed understanding of sex differences will help people make sense of *why* this has happened, how biology may have provided some of the horsepower that has drawn (or pushed) the wagon of social and economic inequality between men and women.

Descriptions, not Prescriptions

Because it is so important, let us again emphasize the distinction between genetic determinism and genetic influence. We human beings are remarkably emancipated from our genes. Our genetic heritage whispers within us; it does not shout. It makes suggestions; it does not issue orders. At the same time, the fact that we are capable of defining and determining our behavior—and thus our present and future—does not necessarily imply that we *must* defy our biology and our past.

The more we understand sex, the more choices we face. For example, a knowledge of the impossibility of unisex babies allows parents to ask such questions as What kind of boys and girls, men and women, do we wish to raise? With what social values? And how does knowledge of maleness and femaleness help us understand our own inclinations and relationships?

As members of a literate society, we feel obligated to educate our children. What should schools tell them about sex differences? What advice is salient enough to be meaningful yet general enough to allow for individual differences and choices? We believe that the basic biology of male–female differences belongs in every curriculum because it is true and because it can encourage personal liberation. Why not expand sex education to include discussions not only of sexual intercourse but also of social intercourse, exploring ways to understand and make sense out of the differences—behavioral no less than anatomical—between men and women? The birds and the bees deserve to be more than a metaphor for sex education; they deserve to be an integral part of it. After all, the social and evolutionary aspects of sex are far more challenging than the plumbing.

Although we are confident that biology generates sex differences we are leery of deriving societal *pre*scriptions from scientific *de*scriptions, even the evolutionary ones we so fervently espouse. A description of differences should never be misunderstood as a prescription for how things ought to be.

No one has yet glimpsed the precise biological limits of the human species, and in the face of such uncertainty, any biologically based recommendations must be leavened with a heavy dose of humility. To be sure, there are plenty of ways in which the biology of sex differences *could* inform social engineering. Would assertiveness training for women and nurturance training for men make for a better world? We suspect so, but frankly we do not know.

If pushed by would-be policy makers, we would be inclined to come down on the side of making weaker and fewer biological assumptions rather than stronger and more plentiful ones, to err—if anything—on the side of overestimating the role of social factors and underestimating that of the biological. This recommendation may surprise our readers, given the approach taken in this book and our confidence that many human sex differences are in fact a result of evolutionary biology.

Nothing, for example, says that women and men are biologically destined to follow different vocational directions. We cannot think of a single profession that requires aptitudes characteristic of only one sex. Brain surgeons need fine motor skill (a female advantage) but also good spatial relations (a male advantage). Truck drivers can make good use of the male asset of upper body strength, at least when loading and unloading their vehicles, but everyone would be better served if their behavior when behind the wheel were less aggressive and more female-like.

Differences at Work

In some ways, men have advantages over women in the conventional workplace, since they are generally more comfortable with competitiveness and with hierarchy. Thus, even though women are more relationship oriented, men may find work relationships more congenial to their temperaments, since these often are not genuine relationships at all but, rather, interactions of convenience and power. However, women may have a special talent for integrative problem solving—for perceiving the whole of a situation—and, especially, a sensitivity to its interpersonal dimensions. Women might have an advantage over men in industries based on service, such as medicine, law, or government, because they are less constricted by competitive worries and more inclined toward seeking cooperative win-win solutions. But certainly there are men who are brilliant and compassionate social engineers and women who are corporate tigers. Our value judgment is only that individuals should have the right to participate fully in society, not that any particular pattern should be enshrined in social traditions.

For women, the problem frequently is less fear of failure than fear of success, since "success," as typically defined by males, tends to be rather unfeminine (as defined by both sexes). Is this a failing in society? In men? Is it a failing in women if they often do not want success as defined by men? Or isn't it also possible that women really do want success, just as men do, and that the argument that women and men define success differently merely serves to paper over legitimate outrage at unequal salaries and workplace advancement?

Whereas men take "aggressive politicking" at work as the norm and expect competition and conflict, many women are stunned by the lack of cooperation and support they receive in the workplace. Women who make it to a company's higher echelons are often surprised to discover that it *is* lonely at the top and that being in charge does not mean being universally liked. Moreover, female strategies and reactions to stress, which can include involuntary tearfulness, may be considered "weak," whereas male strategies such as blustering, bullying, or anger are often taken as a sign of strength, and accorded respect.

Women, by and large, are unprepared for aggressive conflict, aided by neither biology nor cultural training. Judith specializes in helping women physicians deal with stress. She has seen many excellent female doctors sink into despair or depression when faced with, for example, a direct personal challenge brought about by staff infighting, adversarial contract negotiations, or pressure from a clinic manager to pack more patients into an already overcrowded schedule. Once wounded, many struggle to recover the spunk they once had but instead tend to become more passive and distanced from their jobs. In contrast, male physicians confronted with such challenges often go through a denial phase, characterized by outward indifference, then become belligerent, after which they must work to become more composed and restrained.

Making Sense of It All

For all our confidence about the evolutionary biology of sex differences, many questions about men and women remain unanswered. Biology sets certain limits within which everyone must live, but these are quite wide indeed. Many women, for example, pursue short-term sexual escapades; many men treasure lifelong monogamy. There are women who are aggressive, even violent. There are men who are especially cooperative, peaceable, and nurturant.

Making sense of it all may never be possible. But understanding what drives each of us as individuals and as partners can help us unravel the mysteries inherent in both our personal lives and our relationships. From the moment we enter this world until the moment we leave it, our lives are inextricably linked with those of others, even if we—or they—sometimes wish it weren't so. Human relationships range from the positive, framed by love, commitment, support, and trust, to the decidedly negative, characterized by jealousy, domination, distrust, and violence.

We hope this book provides new insights into how conflict as well as intimacy arise within a relationship, and how the differing needs and desires of men and women can be better understood and therefore better balanced. As we work to improve our relationships—with parents, siblings, friends, lovers, children—it helps to understand why we are attracted to the people we are, why raising children free of gender expectations can be so hard, why we may lust for someone yet never wish to marry him or her, why men want more sex with less talk and women want the opposite, and why—even in the most egalitarian relationships—mothers tend to be more nurturing than fathers toward their children.

But for all the emotional anguish sex-based relationships can cause, human existence would be far less pleasurable and exciting without them; in any event, maleness and femaleness are ineradicably and inextricably part of the human condition. It seems only logical, therefore, that everyone should strive for a greater understanding of sex differences: what they are, how they came about, and what, if anything, to do about them.

Even so, our biological heritage is only one aspect of who and what we are. We believe, along with Betty Friedan—whose words provide the epigraph for this chapter—that all people are entitled to make choices in their lives, for better or worse, and moreover, that biology can help provide the information needed to make such choices intelligently. Each human being has the power to choose and shape his or her significant relationships; seeking isolation and intimacy, promiscuity or long-term monogamy. Women can pursue short-term sexual escapades, and men can embrace lifelong fidelity; men can act more like women, and vice versa. We can create cultures that enhance or diminish biological distinctions and laws that are gender blind, gender friendly, or that struggle to counter biologically based inclinations. We believe that women and men will maximize their power to choose when they have maximized their understanding of who they are, that is, who is making those choices—or, in Reinhold Neibuhr's terms, when they have acquired the wisdom to know the difference between what cannot be changed and what can. We believe, further, that wisdom—born of biological knowledge and leavened with social and ethical insight—will provide the power to choose.

Endnotes

Because *Making Sense of Sex* is intended for a general audience, we have carefully cited all quotations and research studies specifically described in the text but have omitted references that are generally well known to the scientific community.

Chapter 1: Differences

p. 4: Delbert Thiessen: "Nature walks through us"—Delbert Thiessen, *Bittersweet Destiny* (New Brunswick, N.J.: Transaction Books, 1996), 3.

p. 5: Variations in sperm number in human ejaculate—R. R. Baker and M. A. Bellis, "Number of Sperm in Human Ejaculates Varies in Accordance with Sperm Competition Theory," *Animal Behaviour* 37 (1989): 867–869; see also R. R. Baker, *The Evolution of Sperm Competition* (New York: Oxford University Press, 1996).

Chapter 2: Biology

p. 12: James Thurber and E. B. White: "Is sex necessary?"—James Thurber and E. B. White, *Is Sex Necessary?* (New York: Harper & Brothers, 1929).

p. 14: Sex as a way of outwitting parasites—W. D. Hamilton, R. Axelrod, and R. Tanese, "Sexual Reproduction as an Adaptation to Resist Parasites (a Review)," *Proceedings of the National Academy of Sciences of the USA* 87 (1990): 3566–3573.

p. 14: The costs of sex—The best "classic" source is George C. Williams, *Sex and Evolution* (Princeton, N.J.: Princeton University Press, 1975); for a more recent and accessible account, see Matt Ridley, *The Red Queen: Sex and the Evolution of Human Nature* (New York: Macmillan, 1994).

p. 15: Arnold, the Long-Necked Preposterous—Shel Silverstein, *Don't Bump the Glump!* (New York: Simon & Schuster, 1964).

p. 17: Female fireflies—James Lloyd, "Aggressive Mimicry in *Photuris* firefly *Femmes Fatales,*" *Science* 149 (1965): 653–654.

p. 18: Courting frogs eaten by bats—Michael J. Ryan, "Sexual Selection and Communication in Frogs," *Trends in Ecology and Evolution* 6 (1991): 351–355; see also Ryan's summary in his book *The Tungara Fro,* (Chicago: University of Chicago Press, 1985).

p. 20: Computer modeling of the origin of gamete differences—G. Parker, R. R. Baker, and V. Smith, "The Origin and Evolution of Gamete Dimorphism and the Male–Female Phenomenon," *Journal of Theoretical Biology* 36 (1972): 529–552; for a more recent and popularized account of the possibilities of such modeling, see Richard Dawkins, *The Blind Watchmaker* (New York: Norton, 1986).

p. 21: Parental investment theory—R. L. Trivers, "Parental Investment and Sexual Selection," in *Sexual Selection and the Descent of Man, 1871–1971,* ed. B. Campbell (Chicago: Aldine, 1972); see also T. H. Clutton-Brock and A. C. J. Vincent, "Sexual Selection and the Potential Reproductive Rates of Males and Females," *Nature* 351 (1991): 58–60.

p. 22: Elephant seal mating system—B. J. LeBoeuf, "Male–Male Competition and Reproductive Success in Elephant Seals," *American Zoologist* 14 (1974): 163–176; there is also a good general review in T. H. Clutton-Brock, *The Evolution of Parental Care* (Princeton, N.J.: Princeton University Press, 1991).

p. 26: Damselfly penis—Jonathan K. Waage, "Dual Function of the Damselfly Penis: Sperm Removal and Transfer," *Science* 203 (1979): 916–918; see also R. L. Smith, ed., *Sperm Competition and the Evolution of Animal Mating Systems* (New York: Academic Press, 1984).

p. 27: Shark penis—William G. Eberhard, *Sexual Selection and Animal Genitalia* (Cambridge, Mass.: Harvard University Press, 1985).

p. 27: "Homosexual rape" in parasitic worms—H. Abele and S. Gilchrist, "Homosexual Rape and Sexual Selection in Acanthocephalan Worms," *Science* 197 (1977): 81–83.

p. 27: Genital size and mating systems in the great apes—Roger V. Short, "Sexual Selection and Its Component Parts, Somatic and Genital Selection, as

Illustrated by Man and the Great Apes," *Advances in the Study of Behaviour* 9 (1979): 131–158.

p. 28: Role reversal among pipefishes—A. Vincent et al., "Pipefishes and Seahorses: Are They All Sex Role Reversed?" *Trends in Ecology and Evolution* 7 (1992): 237–241; A. Berglund, G. Rosenqvist, and I. Svensson, "Mate Choice, Fecundity, and Sexual Dimorphism in Two Pipefish Species (Syngnathidae)," *Behavioral Ecology and Sociobiology* 19 (1986): 301–307.

p. 29: Polyandry among jacanas—Donald Jenni and B. Betts, "Sex Differences in Nest Construction, Incubation, and Parental Behaviour in the Polyandrous American Jacana," *Animal Behaviour* 26 (1978): 207–218; see also Jared Diamond, "Reversal of Fortune," *Discover* 13 (1992): 70–76.

p. 29: Jean-Henri Fabre: "The male is underneath. . . ."—J.-Henri Fabre, *The Life and Love of the Insect*, trans. Alexander Teixeira de Mattos (London: Black, 1914).

p. 30: "Battle of the sexes" in fruit flies—William Rice, "Sexually Antagonistic Male Adaptation Triggered by Experimental Arrest of Female Evolution," *Nature* 381 (1996): 232–234.

p. 31: Sexual selection—Charles Darwin, *The Descent of Man and Selection in Relation to Sex* (London: Murray, 1871).

p. 32: Tail length in African widowbirds—Malte Andersson, "Female Choice Selects for Extreme Tail Length in a Widowbird," *Nature* 299 (1982): 818–820.

p. 32: Forked tails in barn swallows—Anders P. Møller, "Female Choice Selects for Male Sexual Tail Ornaments in the Monogamous Swallow," *Nature* 322 (1988): 640–642.

p. 33: Disease resistance and plumage brightness—W. D. Hamilton and M. Zuk, "Heritable True Fitness and Bright Birds: A Role for Parasites?" *Science* 218 (1982): 384–387.

p. 33: Damselfish courtship—David P. Barash, "Predictive Sociobiology: Mate Selection in Damselfishes and Brood Defense in White-Crowned Sparrows," in *Sociobiology: Beyond Nature/Nurture?* ed. G. Barlow and J. Silverberg (Boulder, Colo.: Westview Press, 1980).

p. 33: Deceptive courtship in pied flycatchers—R. Alatalo et al., "The Conflict Between Male Polygamy and Female Monogamy: The Case of the Pied Flycatcher *Ficedula hypoleuca*," *The American Naturalist* 117 (1981): 738–753; see also H. Temrin and A. Arak, "Polyterritoriality and Deception in Passerine Birds," *Trends in Ecology and Evolution* 4 (1989): 106–108.

p. 34: Extra-pair copulations in red-winged blackbirds—Elizabeth Gray, "Extra-Pair Copulations in Red-Winged Blackbirds (*Agelaius phoeniceus*)" (Ph.D. diss., University of Washington, 1994).

p. 34: Bonobo (pygmy chimpanzee) behavior—T. Kano, *The Last Ape: Pygmy Chimpanzee Behavior and Ecology*, trans. Evelyn Ono Vineberg (Stanford, Calif.: Stanford University Press, 1992).

Chapter 3: Sex

p. 38: Alfred Kinsey and colleagues: "Among all peoples. . . ."—A. C. Kinsey, W. B. Pomeroy, and C. E. Martin, *Sexual Behavior in the Human Male* (Philadelphia: Saunders, 1948).

p. 40: Coolidge effect in rams—W. Beamer, G. Bermant, and M. Clegg, "Copulatory Behaviour of the Ram, *Ovis aries*. II: Factors Affecting Copulatory Satiation," *Animal Behaviour* 17 (1969): 706–711.

p. 40: Montaigne's observations on his stallion—*The Complete Essays of Montaigne*, trans. D. M. Frame (Stanford, Calif.: Stanford University Press, 1958).

p. 41: Lord Byron: "How the devil. . . ."—George Gordon, Lord Byron, "Don Juan," in *Don Juan and Other Satirical Poems* (New York: Odyssey Press, 1935).

p. 41: Kgatla bigamist: "I find them both. . . ."—I. Schapera, *Married Life in an African Tribe* (London: Faber & Faber, 1940).

p. 41: "If you want. . . ."—M. R. Liebowitz, *The Chemistry of Love* (Boston: Little Brown, 1983).

p. 43: Survey of sexual dissatisfaction—Glenn D. Wilson, *Love and Instinct* (London: Temple Smith, 1981).

p. 43: Donald Symons: "Women *give* sex for love. . . ."—D. Symons, *The Evolution of Human Sexuality* (New York: Oxford University Press, 1979).

p. 43: Alfred Kinsey and colleagues, "Most males can immediately understand. . . ."—A. C. Kinsey, W. B. Pomeroy, and C. E. Martin, *Sexual Behavior in the Human Male* (Philadelphia: Saunders, 1948).

p. 43: Mark Twain: "Now there you have. . . ."—Mark Twain, *Letters from the Earth* (New York: Harper & Row, 1962).

p. 44: Status and reproductive success among Mormons—L. Mealey, "The Relationship between Social Status and Biological Success: A Case Study of the Mormon Religious Hierarchy," *Ethology and Sociobiology* 6 (1985): 249–257.

p. 44: Weston La Barre: "When it comes to polygynous. . . ."; "As my friend. . . ."—Weston La Barre, *The Human Animal* (Chicago: University of Chicago Press, 1954), 117 (polygyny: 112).

p. 45: Reproductive success of polygynous women—W. H. Hern, "Polygyny and Fertility Among the Shipibo of the Peruvian Amazon," *Population*

Studies 46 (1992): 53–65; Lee Cronk, "Wealth, Status, and Reproductive Success Among the Mukogodo of Kenya," *American Anthropologist* 93 (1991): 345–361; see also Peter Bretschneider, "Sociobiological Models of Polygyny: A Critical Review," *Anthropos* 87 (1992): 183–191.

p. 45: Polygyny among red-winged blackbirds—Gordon H. Orians, *Some Adaptations of Marsh-Nesting Blackbirds* (Princeton, N.J.: Princeton University Press, 1980).

p. 47: Alfred Kinsey and colleagues: "Many females consider. . . ."—A. C. Kinsey et al. *Sexual Behavior in the Human Female* (Philadelphia: Saunders, 1953).

p. 47: Rewarding sexual experience—Jane Katz, "Your First Night Together," *Cosmopolitan*, March 1997, 221–225.

p. 47: Alfred Kinsey and colleagues: "Most males find it difficult. . . ."— A. C. Kinsey et al., *Sexual Behavior in the Human Female* (Philadelphia: Saunders, 1953).

p. 49: Sexual arousal as a result of viewing pornographic films—V. Sigusch et al., "Psychosexual Stimulation: Sex Differences." *Journal of Sex Research* 6 (1970): 10–14; D. G. Steele and C. E. Walker, "Male and Female Differences in Reaction to Erotic Stimuli as Related to Sexual Adjustment," *Archives of Sexual Behavior* 3 (1974): 459–470; see also J. Shepher and J. Reisman, "Pornography: A Sociobiological Attempt at Understanding," *Ethology and Sociobiology* 6 (1985): 103–114.

p. 50: Self-reporting of sexual fantasies—Glenn D. Wilson, "Male–Female Differences in Sexual Activity, Enjoyment, and Fantasies," *Personality and Individual Differences* 8 (1987): 125–127; see also Glenn D. Wilson and R. J. Lang, "Sex Differences in Sexual Fantasy Patterns," *Personality and Individual Differences* 2 (1981): 343–346.

p. 52: "A profligate sexuality. . . ."—Michael Segell, "Powerful Men: Are They as Good in the Bedroom as They Are in the Boardroom?" *Cosmopolitan*, February 1995, 147–149.

p. 53: Saul Bellow: "As a carnal artist. . . ."—Saul Bellow, *Humboldt's Gift* (New York: Penguin Books, 1984).

p. 53: Ring dove males reject females—C. J. Erickson and P. G. Zenone, "Courtship Differences in Male Ring Doves: Avoidance of Cuckoldry?" *Science* 192 (1976): 1353–1354.

p. 55: Donald Symons: "Among men. . . ."—D. Symons, *The Evolution of Human Sexuality* (New York: Oxford University Press, 1979).

p. 56: Colette: "If he pretends. . . ."—Colette, *The Vagabond* (New York: Farrar, Straus & Young, 1955).

p. 56: Milan Kundera: "On that fateful day. . . ."—Milan Kundera, *The Book of Laughter and Forgetting* (New York: Penguin Books, 1986).

p. 57: "Extramarital sex. . . ."—L. Wolfe, *Playing Around: Women and Extramarital Sex* (New York: Morrow, 1975).

p. 58: Helene Deutsch: "Our impression is that. . . ."—Helene Deutsch, *The Psychology of Women* (New York: Grune & Stratton, 1945).

p. 58: Duration between initial meeting and onset of sexual intimacy— G. M. Peplau, Z. Rubin, and C. T. Hill, "Sexual Intimacy in Dating Relationships," *Journal of Social Issues* 33 (1977): 86–109.

p. 58: Mangaian sexual practices—D. S. Marshall, "Sexual Behavior on Mangaia," *Human Sexual Behavior*, ed. D. S. Marshall and R. C. Suggs (New York: Basic Books, 1971).

p. 59: Nuptial feeding in scorpion flies—R. Thornhill, "Sexual Selection and Nuptial Feeding Behavior in *Bittacus apicalis* (Insecta: Mercoptera)," *American Naturalist* 110 (1976): 529–548; see also R. Thornhill and J. Alcock, *The Evolution of Insect Mating Systems* (Cambridge, Mass.: Harvard University Press, 1983).

p. 60: Simone de Beauvoir: "From primitive times. . . ."—S. de Beauvoir, *The Second Sex* (New York: Bantam Books, 1961).

p. 60: Gifts from Trobriand Island men to women—B. Malinowski, *The Sexual Life of Savages* (New York: Harcourt, Brace, 1923).

p. 60: "The underlying understanding. . . ."—W. H. Davenport, "Sex in Cross-Cultural Perspective," *Human Sexuality in Four Perspective*, ed. F. A. Beach (Baltimore: Johns Hopkins University Press, 1977).

p. 62: George Orwell: "The result, for a tramp. . . ."—George Orwell, *Down and Out in Paris and London* (New York: Harcourt, Brace & World, 1961).

p. 63: "Primitive woman's sexual drive. . . ."—M. J. Sherfey, *The Nature and Evolution of Female Sexuality* (New York: Random House, 1972).

p. 63: Frank Beach: "Any male who entertains. . . ."—F. A. Beach, "Human Sexuality and Evolution," in *Reproductive Behavior*, eds. W. Montagna and W. A. Sadler (New York: Plenum, 1974).

p. 63: Donald Symons: "The sexually insatiable woman. . . ."—D. Symons, *The Evolution of Human Sexuality* (New York: Oxford University Press, 1979).

p. 63: William Acton: "The majority of women. . . ."—William Acton, *Functions and Disorders of the Reproductive Organs in Childhood, Youth, Adult Age, and Advanced Life*, 5th ed. (London: Churchill, 1871).

p. 65: "Orgasm depends on being in love. . . ."—C. Tavris and Susan Sadd, *The Redbook Report on Female Sexuality* (New York: Delacorte Press, 1977).

p. 66: Sexual satisfaction and nonsexual closeness—Seymour Fisher, *The Female Orgasm: Psychology, Physiology, Fantasy* (New York: Basic Books, 1973).

p. 67: Burley hypothesis for the evolution of concealed ovulation—N. Burley, "The Evolution of Concealed Ovulation," *The American Naturalist* 114 (1979): 835–838; see also Jared Diamond, "Sex and the Female Agenda," *Discover* 14 (1993): 86–94.

p. 68: Jane Lancaster: "What would happen. . . ."—Jane B. Lancaster, "Sex Roles in Primate Societies," in *Sex Differences*, ed. M. S. Teitelbaum (New York: Doubleday, 1977).

p. 68: Concealed ovulation as a strategy for obtaining male parental investment—R. D. Alexander and K. Noonan, "Concealment of Ovulation, Parental Care and Human Social Evolution," in *Evolutionary Biology and Human Social Behavior*, eds. N. Chagnon and W. Irons (North Scituate, Mass.: Duxbury Press, 1979).

p. 70: Grandmother hypothesis—Ann Gibbons, "Ideas on Human Origins Evolve at Anthropology Gathering," *Science* 276 (1997): 535–536.

p. 71: Donald Symons's suggestion that rampant sexuality is not unique to gays—D. Symons, *The Evolution of Human Sexuality* (New York: Oxford University Press, 1979).

p. 72: Philip Blumstein and Pepper Schwartz: "Gay men have. . . ."—Philip Blumstein and Pepper Schwartz, *American Couples* (New York: Morrow, 1983).

p. 73: What men and women want in a mate—D. M. Buss, "Sex Differences in Human Mate Preferences: Evolutionary Hypotheses Tested in Thirty-Seven Cultures," *Behavioral and Brain Sciences* 12 (1989): 1–49; see also D. M. Buss, *The Evolution of Desire* (New York: Basic Books, 1994).

p. 73: Sex differences in personal ads—D. Thiessen, R. K. Young, and R. Burroughs, "Lonely Hearts Advertisements Reflect Sexually Dimorphic Mating Strategies," *Ethology and Sociobiology* 14 (1993): 209–229.

p. 74: Research on hypothetical anonymous sex—D. Symons and L. Ellis, "Male–Female Differences in the Desire to Have Intercourse with an Anonymous New Partner," in *Sociobiology of Reproductive Strategies in Animals and Humans*, eds. A. Rasa, C. Vogel, and E. Voland (London: Croom Helm, 1988).

p. 75: Gender gap in situations that trigger sexual jealousy—D. M. Buss et al., "Sex Differences in Jealousy: Evolution, Physiology, and Psychology," *Psychological Science* 3 (1992): 251–255.

p. 77: Dorothy Parker: "Woman wants monogamy. . . ."—Dorothy Parker, "General Review of the Sex Situation," *The Collected Poetry of Dorothy Parker* (New York: Modern Library, 1936).

Chapter 4: Violence

p. 79: Murderous male primates—Richard Wrangham and Dale Peterson, *Demonic Males: Apes and the Origins of Human Violence* (Boston: Houghton Mifflin, 1996).

p. 81: Campbell analysis of the gender gap in aggression—Anne Campbell, *Men, Women, and Aggression* (New York: Basic Books, 1993).

p. 82: Aggressive neglect among male birds—S. D. Ripley, "Aggressive Neglect as a Factor in Interspecific Competition in Birds," *Auk* 78 (1961): 366–371.

p. 82: Sex change in blue-headed wrasses—D. R. Robertson, "Social Control of Sex Reversal in a Coral-Reef Fish," *Science* 177 (1972): 1007–1009.

p. 83: Aggressive politicking among male chimpanzees—F. B. M. de Waal, *Peacemaking Among Primates* (Cambridge, Mass.: Harvard University Press, 1989).

p. 83: Coalitions among chimpanzees—F. de Waal, "Sex Differences in the Formation of Coalitions Among Chimpanzees," *Ethology and Sociobiology* 5 (1984): 239–255.

p. 83: Female–female competition among animals—S. Wasser, ed., *Social Behavior of Female Vertebrates* (New York: Academic Press, 1983).

p. 83: Sarah Hrdy: "Consider . . . such phenomena. . . ."—Sarah Blaffer Hrdy, *The Woman That Never Evolved* (Cambridge, Mass.: Harvard University Press, 1981).

p. 84: Male–female ratio of violent crime—L. Ellis and P. D. Coontz, "Androgens, Brain Functioning and Criminality: The Neurohormonal Foundations of Antisociality," in *Crime in Biological, Social, and Moral Contexts*, eds. L. Ellis and H. Hoffman (New York: Praeger, 1990).

p. 85: Testosterone levels among female prison inmates—J. M. Dabbs et al., "Saliva Testosterone and Criminal Violence in Young Adult Prison Inmates," *Psychosomatic Medicine* 49 (1987): 174–182.

p. 85: Testosterone levels and criminal records—J. M. Dabbs et al., "Saliva Testosterone and Criminal Violence Among Women." *Personality and Individual Differences* 9 (1988): 269–275.

p. 86: Martin Daly and Margo Wilson: "There is no known human society. . . ."—M. Daly and M. Wilson, *Homicide* (Hawthorne, N.Y.: Aldine, 1988).

p. 87: Number of convicted murderers during 1995—*FBI Uniform Crime Reports for the United States* (Washington, D.C.: U.S. Department of Justice, 1996).

p. 88: Reproductive success and wealth of Yomut Turkmen—W. Irons, "Cultural and Biological Success," in *Evolutionary Biology and Human Social Behavior: An Anthropological Perspective*, eds. N. A. Chagnon and W. Irons (North Scituate, Mass.: Duxbury Press, 1979).

p. 88: Power as a predictor of harem size—L. Betzig, *Despotism and Differential Reproduction: A Darwinian View of History* (Hawthorne, N.Y.: Aldine, 1986).

p. 88: Number of potential conceptions among French Canadian men— D. Perusse, "Cultural and Reproductive Success in Industrial Societies: Testing the Relationship at the Proximate and Ultimate Levels," *Behavioral and Brain Sciences* 16 (1993): 267–284.

p. 89: Yanomamö violence—N. Chagnon, "Life Histories, Blood Revenge, and Warfare in a Tribal Population," *Science* 239 (1988): 985–992.

p. 89: Reproductive success in the Yanomamö—N. Chagnon, "Is Reproductive Success Equal in Egalitarian Societies?" in *Evolutionary Biology and Human Social Behavior: An Anthropological Perspective*, eds. N. A. Chagnon and W. Irons (North Scituate, Mass.: Duxbury Press, 1979).

p. 89: Shinbone—N. Chagnon, *Yanomamö: The Fierce People* (New York: Holt, Rinehart & Winston, 1983).

p. 90: *Kepu* among the Dani people—P. Matthiessen, *Under the Mountain Wall: A Chronicle of Two Seasons in the Stone Age* (New York: Viking Press, 1962).

p. 90: "Mexicans, and I think everyone. . . ."—Oscar Lewis, *The Children of Sanchez: Autobiography of a Mexican Family* (New York: Random House, 1961).

p. 91: Edward O. Wilson: "My worst difficulties. . . ."—Edward O. Wilson, *Naturalist* (Washington, D.C.: Island Press/Shearwater Books, 1995), 54–55.

p. 91: Marvin Wolfgang's research—M. Wolfgang, *Patterns in Criminal Homicide* (Philadelphia: University of Pennsylvania Press, 1958).

p. 92: Sexual behavior of squirrel monkeys—D. W. Ploog, J. Blitz, and F. Ploog, "Studies on Social and Sexual Behavior of the Squirrel Monkey (*Saimiri sciureus*)," *Folia Primatologica* 1 (1963): 29–66.

p. 92: Aggression and nitric oxide—R. J. Nelson et al., "Behavioral Abnormalities in Male Mice Lacking Neuronal Nitric Oxide Synthase," *Nature* 378 (1995): 383–386.

p. 92: Proximity of sex and aggression regions in the brain—Paul MacLean, "New Findings Relevant to the Evolution of Psychosexual Functions of the Brain," *Journal of Nervous and Mental Disease* 135 (1962): 289–301.

p. 92: Phallic symbolism of power, dominance, and threat—I. Eibl-Eibesfeldt, *Love and Hate* (New York: Viking Press, 1971).

p. 93: Homosexual rape in prisons—A. J. Davis, "Sexual Assaults in the Philadelphia Prison System," in *The Sexual Scene*, eds. J. H. Gagnon and W. Simon (New Brunswick, N.J.: Transaction Books; Hawthorne, N.Y.: Aldine, 1970); see also C. A. Saum et al., "Sex in Prison," *Prison Journal* 75 (1995): 413–431.

p. 93: Rape among mallard ducks—David P. Barash, "Sociobiology of Rape in Mallards (*Anas platyrhynchos*): Responses of the Mated Male," *Science* 197 (1977): 788–789.

p. 94: Rape among various animals—Numerous references are to be found in J. M. G. van der Dennen, *The Nature of the Sexes* (Groningen, Netherlands: Origin Press, 1992).

p. 94: Rape among chimpanzees—C. E. G. Tutin and R. McGinnis, "Chimpanzee Reproduction in the Wild," in *Reproductive Biology of the Great Apes*, ed. C. E. Graham (New York: Academic Press, 1981).

p. 94: Rape among orangutans—B. M. F. Galdikas, "Orangutan Adaptation at Tanjung Puting Reserve: Mating and Ecology," in *The Great Apes*, eds. D. A. Hamburg and E. McCown (Menlo Park, Calif.: Benjamin-Cummings, 1979).

p. 94: Rape as violence against women—Susan Brownmiller, *Against Our Will: Men, Women, and Rape* (New York: Simon and Schuster, 1975).

p. 94: "In terms of the. . . ."—S. Griffin, *Rape: The Power of Consciousness* (New York: Harper & Row, 1979).

p. 95: Sexual intercourse as the most troublesome form of rape—T. W. McCaheill, L. C. Meyer, and A. M. Fischman, *The Aftermath of Rape* (Lexington, Mass.: Heath, 1979).

p. 96: Marvin Wolfgang's "subculture of violence"—M. Wolfgang and F. Ferracuti, *The Subculture of Violence* (London: Tavistock, 1967).

p. 96: At its diseased heart. . . . —R. Rhodes, "Why Do Men Rape?" *Playboy*, 1981, 112.

p. 96: Twelvefold difference in rape rate—J. M. MacDonald, *Rape Offenders and Their Victims* (Springfield, Ill.: Thomas, 1971).

p. 96: Rape probabilities in cities, suburbs, and so forth—M. S. Eisenhower, *To Establish Justice, to Insure Domestic Tranquility: Final Report of the National Commission of Causes and Prevention of Violence* (Washington, D.C.: U.S. Government Printing Office, 1969).

p. 96: Rape as a function of social class in Denmark—K. Svalastoga, "Rape and Social Structure," *Pacific Sociological Review* 5 (1961): 48–53.

p. 96: India and Gambia, fear of being raped—R. Thornhill and N. W. Thornhill, "Human Rape: The Strengths of the Evolutionary Perspective," in *Sociobiology and Psychology: Ideas, Issues, and Applications*, eds. C. Crawford, M. Smith, and D. Krebs (Hillsdale, N.J.: Erlbaum, 1987).

p. 97: Evolutionary biology of human rape—R. Thornhill and N. W. Thornhill, "Human Rape: The Strengths of the Evolutionary Perspective," in *Sociobiology and Psychology: Ideas, Issues, and Applications*, eds. C. Crawford, M. Smith, and D. Krebs (Hillsdale, N.J.: Erlbaum, 1987); R. Thornhill and N. W. Thornhill, "The Evolutionary Psychology of Men's Coercive Sexuality," *Behavioral and Brain Sciences* 15 (1992): 363–421; see also L. Ellis, *Theories of Rape: Inquiries into the Causes of Sexual Aggression* (New York: Hemisphere, 1989).

p. 97: Gusii bride-wealth payments—R. A. LeVine, "Gusii Sex Offenses: Study in Social Control," in *Forcible Rape: The Crime, the Victim, and the Offender*, eds. D. Chappell, R. Geis, and G. Geis (New York: Columbia University Press, 1977).

p. 97: Low self-esteem of rapists—N. A. Groth and H. J. Birnbaum, *Men Who Rape* (New York: Plenum, 1979).

p. 98: Marriage exonerates a rapist—Calvin Sims, "Justice in Peru: Rape Victim is Pressed to Marry Attacker," *New York Times*, March 8, 1997, A1, A8.

p. 98: Inbreeding-avoidance hypothesis for the incest taboo—E. Westermarck, *The History of Human Marriage* (New York: Macmillan, 1891).

p. 99: Female "rape" of men—B. Malinowski, *The Sexual Life of Savages in Northwestern Melanesia* (New York: Halcyon Press, 1929).

p. 101: Adrienne Zihlman and Nancy Tanner: "Females preferred to associate. . . ."—Adrienne Zihlman and Nancy Tanner, "Women in Evolution, Part I: Innovation and Selection in Human Origins," *Signs* 1 (1976): 585–608.

p. 102: Alfred Kinsey and colleagues: "Wives, at every social level, . . ."—A. C. Kinsey, W. B. Pomeroy, and C. E. Martin, *Sexual Behavior in the Human Male* (Philadelphia: Saunders, 1948).

p. 103: Men cite adultery as cause of divorce more than do women—M. Daly, M. Wilson, and S. Wedhorst, "Male Sexual Jealousy," *Ethology and Sociobiology* 3 (1987): 1–27.

p. 103: Infidelity as a cause of divorce—Laura Betzig, "Causes of Conjugal Dissolution: A Cross Cultural Study," *Current Anthropology* 30 (1989): 654–676.

p. 103: Mountain bluebird "adultery"—D. P. Barash, "The Male Response to Apparent Female Adultery in the Mountain Bluebird, *Sialia currucoides*: An Evolutionary Interpretation," *The American Naturalist* 110 (1976): 1097–1101.

p. 104: Seclusion of high-status women in northern India—M. Dickemann, "Paternal Confidence and Dowry Competition: A Biocultural Analysis of Purdah," in *Natural Selection and Social Behavior: Recent Research and New Theory,* eds. R. D. Alexander and D. Tinkle (Concord, Mass.: Chiron, 1981).

p. 105: Reasons for spousal homicide—M. Daly and M. Wilson, *Homicide* (Hawthorne, N.Y.: Aldine, 1988).

p. 106: Men kill women out of sexual jealousy more often than women kill men—M. Daly, M. Wilson, and S. J. Weghorst, "Male Sexual Jealousy," *Ethology and Sociobiology* 3 (1982): 11–27.

p. 106: "She said that since. . . ."—C. A. Carlson, "Intrafamilial Homicide: A Sociobiological Perspective" (B.Sc. thesis, McMaster University, 1984), cited in M. Daly and M. Wilson, *Homicide* (Hawthorne, N.Y.: Aldine, 1988).

p. 107: "You see, we were always arguing. . . ."—P. D. Chambos, *Marital Violence: A Study of Interspouse Homicide* (San Francisco: R & E Research Associates, 1978).

Chapter 5: Parenting

p. 109: Women do more child care than men, cross-culturally—M. Katz and M. Konner, "The Role of the Father: An Anthropological Perspective," *The Role of the Father in Child Development,* ed. M. Lamb (New York: Wiley, 1981).

p. 110: Paternal behavior omitted—M. Bornstein, ed., *Cultural Approaches to Parenting* (Hillsdale, N.J.: Erlbaum, 1991).

p. 110: Ye'kwana parental care—R. Hames, "The Allocation of Parental Care Among the Ye'kwana," in *Human Reproductive Behavior: A Darwinian Perspective,* ed. L. Betzig, M. Borgerhoff Mulder, and P. Turke (Cambridge: Cambridge University Press, 1988).

p. 111: "When the child. . . ."—B. S. Hewlett, "Sexual Selection and Paternal Investment Among Aka Pygmies," in *Human Reproductive Behavior, A Darwinian Perspective,* ed. L. Betzig, M. Borgerhoff Mulder, and P. Turke (Cambridge: Cambridge University Press, 1988).

p. 111: Survey of rural and nontechnological societies—Beatrice Whiting and John Whiting, *Children of Six Cultures: A Psychocultural Analysis* (Cambridge, Mass.: Harvard University Press, 1975).

p. 113: Shulamith Firestone: "The heart of woman's oppression. . . ."—Shulamith Firestone, *The Dialectic of Sex* (New York: Morrow, 1970).

p. 114: Women want to be the primary caregivers—N. Radin, "Primary Caregiving and Role-Sharing Fathers," in *Nontraditional Families: Parenting and Child Development* (Hillsdale, N.J.: Erlbaum, 1982); see also T. G. Power,

"Mother- and Father-Infant Play: A Developmental Analysis," *Child Development* 56 (1985): 1514–1524.

p. 116: Melford Spiro's kibbutz findings—Melford Spiro, *Gender and Culture: Kibbutz Women Revisited* (Durham, N.C.: Duke University Press, 1979).

p. 116: Children in rural and traditional cultures—Beatrice Whiting and John Whiting, *Children of Six Cultures: A Psychocultural Analysis* (Cambridge, Mass.: Harvard University Press, 1975).

p. 117: Second-time fathers are not more involved with their infants—E. Shapiro, "Transition to Parenthood in Adult and Family Development" (Ph.D. diss., University of Massachusetts, 1979).

p. 117: Recognition of infants' facial expressions—W. A. Babchuck, R. B. Hames, and R. A. Thompson, "Sex Differences in the Recognition of Infant Facial Expressions of Emotion: The Primary Caretaker Hypothesis," *Ethology and Sociobiology* 6 (1983): 89–102.

p. 117: Parents' responses to infants' distress—A. R. Weisenfeld, C. Z. Malatesta, and L. L. DeLoach, "Differential Parental Response to Familiar and Unfamiliar Infant Distress Signals," *Infant Behavior and Development* 4 (1981): 281–295; see also M. R. Adelsberg, "The Effects of the Sex of the Parent, the Sex of the Infant and the Type of Family on the Parent's Attunement to Their Distressed Infant and Play and Soothing Style Utilized" (Ph.D. diss., Long Island University, 1989).

p. 117: Spandrels—S. J. Gould, "Evolution: The Pleasures of Pluralism," *New York Review of Books* XLIV (1997): 47–52.

p. 118: Maternal sensitivity and compassion—Sara Ruddick, *Maternal Thinking: Toward a Politics of Peace* (Boston: Beacon Press, 1995).

p. 118: Play with fathers—M. E. Lamb, "Interactions Between Two-Year-Olds and Their Mothers and Fathers," *Psychological Reports* 38 (1976): 447–450.

p. 118: Alice Rossi's case study of Stuart—Alice S. Rossi, "Gender and Parenthood," *Gender and the Life Course*, ed. A. S. Rossi (Hawthorne, N.Y.: Aldine, 1985).

p. 120: Parental care among gladiator frogs—William E. Duellman, "Reproductive Strategies of Frogs," *Scientific American* 267 (1992): 80–87.

p. 120: Lactation in male fruit bats—C. M. Francis, E. L. P. Anthony, J. A. Branton, and T. H. Kunz, "Lactation in Male Fruit Bats," *Nature* 367 (1994): 691–692.

p. 121: August Strindberg: "I know of nothing so ludicrous. . . ."—August Strindberg, *The Father*, trans. M. Meyer (New York: Modern Library, 1966).

p. 122: Paternity confidence in water bugs—R. Smith, "Repeated Copulation and Sperm Precedence: Paternity Assurance for a Brooding Male Water Bug," *Science* 205 (1979): 1029–1031.

p. 123: Reproduction by mates of vasectomized blackbirds—O. Bray, J. Kennelly, and J. Guarino, "Fertility Eggs Produced on Territories of Vasectomized Red-Winged Blackbirds," *The Wilson Bulletin* 87 (1975): 187–195.

p. 124: Male baboons' solicitousness to the young of their prior consorts—H. Klein, "Paternal Care in Free-Living Yellow Baboons" (Ph.D. diss., University of Washington, 1981).

p. 124: Mountain bluebird stepfathers—H. Power, "Mountain Bluebirds: Experimental Evidence Against Altruism," *Science* 189 (1975): 142–143.

p. 124: Dunnock paternity and parental care—N. B. Davies et al., "Paternity and Parental Effort in Dunnocks (*Prunella modularis*): How Good are Male Chick-Feeding Rules?" *Animal Behaviour* 43 (1992): 729–745; see also A. Dixon et al., "Paternal Investment Inversely Related to Degree of Extra-Pair Paternity in the Reed Bunting," *Nature* 371 (1994): 698–700.

p. 125: Cloacal pecking in dunnocks—N. B. Davies, "Polyandry, Cloaca Pecking and Sperm Competiton in Dunnocks," *Nature* 302 (1983): 334–336.

p. 125: Comments on paternal resemblance at the birth of a baby—M. Daly and M. Wilson, "Whom Are Newborn Babies Said to Resemble?" *Ethology and Sociobiology* 3 (1982): 69–78.

p. 126: Paternal resemblance among Mexican babies—J. M. Regalski and S. J. C. Gaulin, "Whom Are Mexican Infants Said to Resemble? Monitoring and Fostering Paternal Confidence in the Yucatan," *Ethology and Sociobiology* 14 (1993): 97–113.

p. 126: Male baboons' tolerance of potential mates' offspring—B. Smuts, *Sex and Friendship in Baboons* (Hawthorne, N.Y.: Aldine, 1985).

p. 126: Male vervet monkeys' deviousness—A. C. Keddy Hector, R. M. Seyfarth, and M. J. Raleigh, "Male Parental Care, Female Choice and the Effect of an Audience in Vervet Monkeys," *Animal Behaviour* 38 (1989): 262–271.

p. 127: Cads and dads—Richard Dawkins, *The Selfish Gene* (New York: Oxford University Press, 1989).

p. 127: Cads and dads in Trinidad—M. K. Flinn, "Paternal Care in a Caribbean Village," in *Father-Child Relations*, ed. B. Hewlett (Hawthorne, N.Y.: Aldine, 1992).

p. 129: Greater risk of children being killed by nonbiological parents—M. Daly and M. Wilson, "Child Abuse and Other Risks of Not Living with Both Parents," *Ethology and Sociobiology* 6 (1985): 197–210.

p. 129: Dian Fossey's observations of gorilla infanticide—Dian Fossey, "The Behaviour of the Mountain Gorilla" (Ph.D. diss., Cambridge University, 1976).

p. 129: Langur infanticide—Sarah B. Hrdy, *The Langurs of Abu* (Cambridge, Mass.: Harvard University Press, 1977); see also Glen Hausfater and S. B.

Hrdy, *Infanticide: Comparative and Evolutionary Perspectives* (Hawthorne, N.Y.: Aldine, 1984).

p. 130: Influence of testosterone on parental behavior in male birds—J. C. Wingfield et al., "The 'Challenge Hypothesis': Theoretical Implications for Patterns of Testosterone Secretion, Mating Systems, and Breeding Strategies," *The American Naturalist* 136 (1990): 829–845.

p. 130: Influence of testosterone on parental behavior in mammals—M. M. West and M. J. Konner, "The Role of the Father: An Anthropological Perspective," in *The Role of the Father in Child Development*, ed. M. E. Lamb (New York: Wiley, 1976); see also Carolyn T. Halpern et al., "Relationships Between Aggression and Pubertal Increases in Testosterone: A Panel Analysis of Adolescent Males," *Social Biology* 40 (1993): 8–24.

p. 130: Role of fosB gene in maternal behavior—J. R. Brown et al., "A Defect in Nurturing in Mice Lacking the Immediate Early Gene fosB," *Cell* 86 (1996): 297–309.

Chapter 6: Childhood

p. 135: Margaret Mead: "There is a long, long road. . . ."—Margaret Mead, *Male and Female* (New York: Morrow, 1948).

p. 136: Rhesus monkeys reared in isolation—H. Harlow, "Sexual Behavior in the Rhesus Monkey," in *Sex and Behavior*, ed. F. A. Beach (New York: Wiley, 1965).

p. 137: Elephant seal super-weaners—B. J. LeBoeuf, R. J. Whiting, and R. Gantt, "Perinatal Behavior of Northern Elephant Seal Females and Their Young," *Behaviour* 43 (1972): 121–156.

p. 137: Boy–girl differences in risk taking—H. J. Ginsburg and S. M. Miller, "Sex Differences in Children's Risk Taking," *Child Development* 53 (1982): 426–428.

p. 137: Coyness of little girls—I. Eibl-Eibesfeldt, *Love and Hate: On the Natural History of Basic Behaviour Patterns* (London: Methuen, 1971).

p. 138: Attraction to images of things versus people—D. McGuinness and J. Symonds, "Sex Differences in Choice Behavior: The Object-Person Dimension," *Perception* 6 (1977): 691–694.

p. 138: Simultaneous viewing of things and people—D. McGuiness, "Sex Differences in Organization, Perception, and Cognition," in *Exploring Sex Differences*, ed. B. Lloyd and J. Archer (London: Academic Press, 1976).

p. 138: Fussiness in infants—Howard A. Moss, "Early Sex Differences and Mother-Infant Interaction," in *Sex Differences in Behavior*, ed. R. Friedman, R. Richart, and R. L. Vande Wiele (New York: Wiley, 1974).

p. 139: Gender roles among the !Kung San—Patricia Draper, "!Kung Women: Contrasts in Sexual Egalitarianism in Foraging and Sedentary Contexts," in *Toward an Anthropology of Women*, ed. R. Reiter (New York: Monthly Review Press, 1975); see also Roberta L. Hall and Patricia Draper, *Male-Female Differences* (New York: Praeger, 1985).

p. 139: Differences in boys' and girls' drawings—S. Feinberg, "Conceptual Content and Spatial Characteristics of Boys' and Girls' Drawings of Fighting and Helping," *Studies in Art Education* 18 (1977): 63–72.

p. 139: Differences in boys' and girls' storytelling—E. G. Pitcher, "An Interest in Persons as an Aspect of Sex Differences in the Early Years," *Genetic Psychology Monographs* 55 (1957): 287–323.

p. 140: Barrie Thorne: "Boys' groups are larger. . . ."—Barrie Thorne, *Gender Play* (New Brunswick, N.J.: Rutgers University Press, 1993).

p. 140: Girls' tendency to seek approval through cooperation—S. Harter, "Mastery, Motivation and the Need for Approval in Older Children and Their Relationship to Social Desirability Response Tendencies," *Developmental Psychology* 11 (1975): 186–196.

p. 141: Cross-cultural nature of same-sex play groups—Z. Rubin, *Children's Friendships* (Cambridge, Mass.: Harvard University Press, 1980).

p. 141: Janet Lever's observations of play patterns—Janet Lever, "Sex Differences in the Games Children Play," *Social Problems* 23 (1976): 478–487; see also J. Lever, "Sex Differences in the Complexity of Children's Play and Games," *American Sociological Review* 43 (1978): 471–483.

p. 141: Evelyn Pitcher and Lynn Schultz's observations of play patterns—Evelyn G. Pitcher and Lynn Hickey Schultz, *Boys and Girls at Play* (New York: Praeger, 1983).

p. 143: "If some mad sociologist. . . ."—T. H. Middleton, "Boys and Girls Together," *Saturday Review of Literature*, May 1980.

p. 144: Beatrice Whiting and Carolyn Edwards's cross-cultural study of child development—Beatrice B. Whiting and Carolyn P. Edwards, *Children of Different Worlds* (Cambridge, Mass.: Harvard University Press, 1988).

p. 144: Lawrence Kohlberg's cognitive-developmental view of socialization—Lawrence Kohlberg, "Stage and Sequence: The Cognitive-Developmental Approach to Socialization," in *Handbook of Socialization Theory and Research*, ed. D. A. Goslin (Chicago: Rand McNally, 1969); see also Lawrence Kohlberg, *The Psychology of Moral Development* (San Francisco: Harper & Row, 1984).

p. 145: Adults' perception of toddlers' fear or anger—J. Condry and S. Condry, "Sex Differences: A Study of the Eye of the Beholder," *Child Development* 47 (1976): 812–819.

p. 145: Rhesus monkey mothers' differential treatment of male and female offspring—G. Mitchell and E. M. Brandt, "Behavioral Differences Related to Experience of Mother and Sex of Infant in the Rhesus Monkey," *Developmental Psychology* 3 (1970): 149.

p. 145: Parents' behavior toward infants depending on infants' sex—J. Z. Rubin, F. J. Provenzano, and Z. Luria, "The Eye of the Beholder: Parents' Views on Sex of Newborns," *American Journal of Orthopsychiatry* 44 (1974): 512–519.

p. 145: Parents' style of conversing with young children—L. Cherry and M. Lewis, "Mothers and Two-Year-Olds: A Study of Sex-Differentiated Aspects of Verbal Interaction," *Developmental Psychology* 12 (1976): 278–282.

p. 146: Experiences of Fulani boys: "At about six years. . . ."—D. F. Lott and B. L. Hart, "Aggressive Domination of Cattle by Fulani Herdsmen and in Relation to Aggression in Fulani Culture and Personality," *Ethos* 5 (1977): 174–186.

p. 146: Disproportionate punishment of boys—G. C. Walters and J. G. Grusex, *Punishment* (San Francisco: Freeman, 1977).

p. 147: Differences between girls' and boys' bedrooms—H. Rheingold and K. Cook, "The Content of Boys' and Girls' Rooms as an Index of Parents' Behavior," *Child Development* 46 (1975): 59–63.

p. 147: Parents' tendency to dichotomize—W. S. Barnes, "Sibling Influences Within Family and School Contexts" (Ph.D. diss., Harvard University, 1984).

p. 148: Boys' lack of attention in caring for babies and younger siblings—P. W. Berman, "Young Children's Responses to Babies: Do They Foreshadow Differences Between Maternal and Paternal Styles?" in *Origins of Nurturance: Developmental, Biolological and Cultural Perspectives on Caregiving*, ed. A. Fogel and G. F. Melson (Hillsdale, N.J.: Erlbaum, 1986).

p. 148: Beatrice Whiting and Carolyn Edwards: "Jubenal, a seven-year-old, . . ."—Beatrice B. Whiting and Carolyn P. Edwards, *Children of Different Worlds* (Cambridge, Mass.: Harvard University Press, 1988).

p. 149: Preferences for male or female offspring—R. L. Trivers and D. E. Willard, "Natural Selection of Parental Ability to Vary the Sex Ratio of Offspring," *Science* 179 (1973): 90–92; see also S. B. Hrdy, "Sex-Biased Parental Investment Among Primates and Other Mammals: A Critical Re-evaluation of the Trivers-Willard Hypothesis," in *Child Abuse and Neglect: Bio-social Dimensions*, ed. R. Gelles and J. Lancaster (Hawthorne, N.Y.: Aldine, 1987).

p. 150: Socioeconomic status and preferred sex of offspring—Mildred Dickemann, "Female Infanticide, Reproductive Strategies, and Social Stratification: A Preliminary Model," in *Evolutonary Biology and Human Social Be-

havior, ed. N. Chagnon and W. Irons (North Scituate, Mass.: Duxbury Press, 1979); see also M. Dickemann, "Phylogenetic Fallacies and Sexual Oppression," *Human Nature* 3 (1992): 71–87.

Chapter 7: Body

p. 155: Basic body differences—S. M. Bailey, "Absolute and Relative Sex Differences in Body Composition," *Sexual Dimorphism in* Homo sapiens: *A Question of Size,* ed. R. L. Hall (New York: Praeger, 1982).

p. 156: Presence of consistently more body fat in girls—F. L. Smoll and R. W. Schutz, "Quantifying Gender Differences in Physical Performance: A Developmental Perspective," *Developmental Psychology* 26 (1990): 360–369.

p. 156: Differences in grip strength—R. R. Montpetit, H. J. Montoye, and L. Laeding, "Grip Strength of School Children, Saginaw, Michigan—1964," *Research Quarterly* 38 (1967): 231–240.

p. 156: Fatness and fertility—R. Z. Van der Spuy, "Nutrition and Reproduction," *Clinical Obstetrics and Gynaecology* 12 (1985): 579–604; see also M. de Souza and D. Metzger, "Reproductive Dysfunction in Amenorrheic Athletes and Anorexic Patients: A Review." *Medicine and Science in Sports and Exercise* 23 (1991): 995–1007.

p. 156: Differences in fat, muscle, and so on—R. Malina, "Quantification of Fat, Muscle, and Bone in Man." *Clinical Orthopaedics* 65 (1969): 9–38.

p. 158: Correlation of dimorphism and degree of polygyny—Richard D. Alexander et al., "Sexual Dimorphisms and Breeding Systems in Pinnipeds, Ungulates, Primates, and Humans," in *Evolutonary Biology and Human Social Behavior,* ed. N. Chagnon and W. Irons (North Scituate, Mass.: Duxbury Press, 1979).

p. 158: Decline in human sexual dimorphism—Henry M. McHenry, "How Big Were Early Hominids?" *Evolutionary Anthropology* 1 (1992): 15–20.

p. 158: Women's preference for tall men—David M. Buss, *The Evolution of Desire* (New York: Basic Books, 1994); see also L. A. Jackson, *Physical Appearance and Gender: Sociobiological and Sociocultural Perspectives* (Albany: State University of New York Press, 1992).

p. 160: Sarah Hrdy: "The virtues of large size. . . ."—Sarah Hrdy, *The Woman That Never Evolved* (Cambridge, Mass.: Harvard University Press, 1981), 24–25.

p. 160: Higher costs of rearing males—T. H. Clutton-Brock, "The Costs of Sex," in *The Differences Between the Sexes,* ed. R. V. Short and E. Balaban (Cambridge: Cambridge University Press, 1994).

p. 162: James Joyce: "Every physical quality. . . ."—James Joyce, *A Portrait of the Artist as a Young Man* (New York: Viking Press, 1956).

p. 164: Ratio of waist to hip measurements in women—D. Singh, "Adaptive Significance of Waist-to-Hip Ratio and Female Physical Attractiveness," *Journal of Personality and Social Psychology* 65 (1993): 293–307.

p. 165: Attractiveness of facial symmetry—R. Thornhill and S. W. Gangestad, "Human Facial Beauty: Averageness, Symmetry, and Parasite Resistance," *Human Nature* 4 (1993): 237–269.

p. 165: Sexual attractiveness of the chimpanzee Flo—Jane Goodall, *In the Shadow of Man* (Boston: Houghton Mifflin, 1971).

p. 165: Hairlessness and skin color preference—P. L. vandenBerghe and P. Frost, "Skin Color Preference, Sexual Dimorphism, and Sexual Selection: A Case of Gene-Culture Co-Evolution?" *Ethnic and Racial Studies* 9 (1986): 87–118.

p. 166: Smaller feet in women—W. D. Ross and R. Ward, "Human Proportionality and Sexual Dimorphism," in *Sexual Dimorphism in* Homo sapiens: *A Question of Size*, ed. R. L. Hall (New York: Praeger, 1982).

p. 166: Preference of men for youthful sexual partners—K. Grammer, "Variations on a Theme: Age-Dependent Mate Selection in Humans," *Behavioral and Brain Sciences* 15 (1992): 100–102.

p. 168: Characteristics women seek in sperm donors—J. E. Scheib, "Sperm Donor Selection and Psychology of Female Mate Choice," *Ethology and Sociobiology* 15 (1994): 113–129; J. E. Scheib, A. Kristiansen, and A. Wara, "A Norwegian Note on 'Sperm Donor Selection and the Psychology of Female Mate Choice'," *Evolution and Human Behavior* 18 (1997): 143–149.

p. 169: Sex differential in longevity—D. L. Wingard, "The Sex Differential in Morbidity, Mortality, and Lifestyle," *Annual Review of Public Health* 5 (1984): 433–458.

p. 170: Effect of castration on longevity—J. B. Hamilton and G. E. Mestler, "Mortality and Survival: Comparison of Eunuchs with Intact Men and Women in a Mentally Retarded Population," *Journal of Gerontology* 24 (1969): 395–411.

p. 170: Differences in abortion rate and survival—J. M. Tanner, *Fetus into Man: Physical Growth from Conception to Maturity* (Cambridge, Mass.: Harvard University Press, 1978).

Chapter 8: Brain

p. 177: Prenatal masculinization in spotted hyenas—L. G. Frank, S. E. Glickman, and I. Powch, "Sexual Dimorphism in the Spotted Hyena," *Journal of Zoology* 221 (1990): 308–313; see also L. G. Frank, S. E. Glickman, and P. Licht, "Fatal Sibling Aggression, Precocial Development and Androgens in Neonatal Spotted Hyenas," *Science* 252 (1991): 702–704.

p. 177: Effect of prenatal testosterone exposure on rhesus monkeys—C. H. Phoenix, "Prenatal Testosterone in the Nonhuman Primate and its Consequences for Behavior," in *Sex Differences in Behavior*, eds. R. C. Friedman, R. M. Richart, and R. L. vande Wiele (New York: John Wiley, 1974).

p. 179: John Money and Anke Ehrhardt: "All control girls. . . ."—John Money and Anke Ehrhardt, *Man & Woman, Boy & Girl* (Baltimore: Johns Hopkins University Press, 1972), 101–102.

p. 179: Influence of CAH on toy preference—S. A. Berenbaum and M. Hines, "Early Androgens are Related to Childhood Sex-Typed Toy Preferences," *Psychological Science* 3 (1992): 203–206.

p. 181: Neurobiology of male–female differences—Robert Pool, *Eve's Rib* (New York: Crown, 1994).

p. 182: Marriage and individuals with AIS—John Money and Anke Ehrhardt, *Man & Woman, Boy & Girl*, (Baltimore: Johns Hopkins University Press, 1972).

p. 182: 5ARDS—J. Imperato-McGinley et al., "Androgens and the Evolution of Male-Gender Identity Among Male Pseudo-Hermaphrodites with 5a-reductase Deficiency," *The New England Journal of Medicine* 300 (1979): 1233–1237; see also Jared Diamond, "Turning a Man," *Discover* 13 (1992): 70–77.

p. 182: Hercule/Herculine Barbin—R. McDougall, *Herculine Barbin: Being the Recently Discovered Memoirs of a Nineteenth-Century French Hermaphrodite* (New York: Pantheon Books, 1980).

p. 183: Penile amputation and the Joan/John case—Milton Diamond, "Some Genetic Considerations in the Development of Sexual Orientation," in *The Development of Sex Differences and Similarities in Behavior*, eds. M. Haug, R. Whalen, C. Aron, and K. Olsen (Boston: Kluwer, 1993); see also Natalie Angier, "Sexual Identity Not Pliable, After All, Report Says," *New York Times* March 14, 1997, A1, A10.

p. 184: Turner's syndrome—E. McCauley et al., "Turner Syndrome: Cognitive Deficits, Affective Discrimination and Behavior Problems," *Child Development* 58 (1987): 464–473.

p. 185: J. R. Urdry: "The first edition. . . ."—J. R. Urdry, *The Social Context of Marriage* (Philadelphia: Lippincott, 1974).

p. 185: Diane Halpern: "At the time. . . ."—Diane Halpern, *Sex Differences in Cognitive Abilities* (Hillsdale, N.J.: Erlbaum, 1986).

p. 185: Alfred Binet's revision of the IQ test—Diane McGuinness, *When Children Don't Learn: Understanding the Biology and Psychology of Learning Disabilities* (New York: Basic Books, 1985).

p. 186: Differences in SAT scores—Camilla P. Benbow and Julian C. Stanley, "Sex Differences in Mathematical Ability: Fact or Artifact?" *Science* 210 (1980): 1262–1264.

p. 186: Larger sample size—Camilla P. Benbow and Julian C. Stanley, "Sex Differences in Mathematical Reasoning Ability: More Facts," *Science* 222 (1983): 1029–1031.

p. 187: Mathematics test scores of boys—Larry V. Hedges and Amy Nowell, "Sex Differences in Mental Test Scores, Variability, and Numbers of High-Scoring Individuals," *Science* 269 (1995): 41–45.

p. 188: Tillie Olsen on the intellectual cost of mothering—Tillie Olsen, *Silences* (New York: Dell, 1978).

p. 191: Alphabet task (identifying shape versus sound)—M. Coltheart, E. Hull, and D. Slater, "Sex Differences in Imagery and Reading," *Nature* 253 (1975): 438–440.

p. 191: Male–female difference in spatial ability—S. Cole-Harding, A. L. Morstad, and J. R. Wilson, "Spatial Ability in Members of Opposite-Sex Twin Pairs," *Behavior Genetics* 18 (1988): 710.

p. 192: Menstrual cycle and cognition—Elizabeth Hampson, "Variations in Sex-Related Cognitive Abilities Across the Menstrual Cycle," *Brain and Cognition* (1990): 26–43.

p. 192: Sex hormones and cognition—J. Imperato-McGinley et al., "Cognitive abilities in Androgen-Insensitive Subjects: Comparison with Control Males and Females form the Same Kindred," *Clinical Endocrinology* 34 (1991): 341–347.

p. 193: Male–female difference in neuron density—Sandra F. Witelson, I. I. Glezer, and D. L. Kigar, "Women Have Greater Density of Neurons in Posterior Temporal Cortex," *Journal of Neuroscience* 15 (1995): 3418–3428.

p. 193: Brain differences in male and female animals—L. Jacobs, "Sexual Selection and the Brain," *Trends in Research in Ecology and Evolution* 11 (1996): 82–86.

p. 193: Male–female difference in size of rat SDN—R. A. Gorski et al., "Evidence for a Morphological Sex Difference Within the Medial Preoptic Area of the Rat Brain," *Brain Research* 148 (1978): 333–346.

p. 194: Testosterone's influence on size of rat SDN—C. D. Jacobson et al., "The Influence of Gonadectomy, Androgen Exposure or a Gonadal Graft in the Neonatal Rat on the Volume of the Sexually Dimorphic Nucleus of the Preoptic Area," *Journal of Neuroscience* 1 (1981): 1142–1147.

p. 194: Human SDN—D. F. Swaab and E. Fliers, "A Sexually Dimorphic Nucleus in the Human Brain," *Science* 228 (1985): 1112–1115.

p. 194: INAH-3 difference in heterosexuals and homosexuals—Simon LeVay, "A Difference in Hypothalamic Structure Between Heterosexual and Homosexual Men," *Science* 253 (1991): 1034–1037; see also Simon LeVay, *The Sexual Brain* (Cambridge, Mass.: MIT Press, 1993).

p. 194: Genetics of sexual orientation—J. M. Bailey and R. C. Pillard, "A Genetic Study of Male Sexual Orientation," *Archives of General Psychiatry* 48 (1991): 1089–1096; and J. M. Bailey et al., "Heritable Factors Influence Sexual Orientation in Women," *Archives of General Psychiatry* 50 (1993): 217–223; Simon LeVay, *Queer Science* (Cambridge, Mass.: MIT Press, 1996).

p. 195: Transsexuality and brain regions—D. F. Swaab, L. J. Gooren, and M. A. Hofman, "Brain Research, Gender, and Sexual Orientation," *Journal of Homosexuality* 28 (1995): 283–301.

p. 195: Male–female differences in brain lateralization—D. Kimura, "Sex Differences in the Brain," *Scientific American* 267 (1993): 118–125.

p. 196: MRI scans of cerebral blood flow—B. A. Shaywitz et al., "Sex Differences in the Functional Organization of the Brain for Language," *Nature* 373 (1995): 607–609.

p. 196: Brain's vulnerability to damage—D. Kimura, "Sex Differences in Cerebral Organization for Speech and Praxic Functions," *Canadian Journal of Psychology* 37 (1983): 19–35.

Chapter 9: The Power to Choose

p. 204: Cynthia Fuchs Epstein: "Deceptive distinctions"—Cynthia Fuchs Epstein, *Deceptive Distinctions* (New Haven, Conn.: Yale University Press, 1988).

p. 205: Carol Gilligan's book—Carol Gilligan, *In a Different Voice* (Cambridge, Mass.: Harvard University Press, 1983).

p. 205: Deborah Tannen's book—Deborah Tannen, *You Just Don't Understand* (New York: Morrow, 1990).

p. 205: Difference feminism—Katha Pollitt, "Are Women Morally Superior to Men?" *The Nation*, December 28, 1992, 799–807.

p. 207: E. O. Wilson: "Genes hold culture on a leash"—Edward O. Wilson, *On Human Nature* (Cambridge, Mass.: Harvard University Press, 1978); see also Charles J. Lumsden and Edward O. Wilson, *Genes, Mind, and Culture* (Cambridge, Mass.: Harvard University Press, 1981).

Index